A WHITE HOUSE OF STONE

A WHITE HOUSE OF STONE

Building America's First Ideal in Architecture

WILLIAM SEALE

THE **WHITE HOUSE** *HISTORICAL ASSOCIATION*

*To all whose labor built this house
and all who labor in it still.*

THE WHITE HOUSE HISTORICAL ASSOCIATION

The White House Historical Association is a nonprofit educational association founded in 1961 for the purpose of enhancing the understanding, appreciation, and enjoyment of the Executive Mansion. All proceeds from the sale of the association's books and products are used to fund the acquisition of historic furnishings and art work for the permanent White House collection, assist in the preservation of public rooms, and further its educational mission.

Produced with the assistance of Furthermore: A Program of the J.M. Kaplan Fund.

BOARD OF DIRECTORS

Frederick J. Ryan Jr., Chairman
John F. W. Rogers, Vice Chairman
James I. McDaniel, Secretary
John T. Behrendt, Treasurer
Stewart D. McLaurin, President

EDITORIAL AND PRODUCTION STAFF

Marcia Mallet Anderson, Vice President of Publishing and Executive Editor
Fiona Griffin, Editorial Director
Lauren A. Zook, Senior Production Manager
Kristin Skinner, Production Manager
Olivia Sledzik, Editorial Specialist
Ann Hofstra Grogg, Consulting Editor

DESIGNER

Pentagram

COPYRIGHT © 2017 WILLIAM SEALE.

All rights reserved under international copyright conventions. No part of this book may be reproduced or utilized in any form or by any means, electronic or mechanical, including photocopying, recording, or by any information storage and retrieval system, without permission in writing from the publisher. Unless otherwise noted, all photographs are copyrighted by the White House Historical Association. Requests for reprint permissions should be addressed to Photo Archivist, White House Historical Association, P.O. Box 27624, Washington, D.C. 20038.

FIRST EDITION
10 9 8 7 6 5 4 3 2 1
LIBRARY OF CONGRESS CONTROL NUMBER: 2017933006
ISBN 978-1-931917-77-3
PRINTED IN ITALY

Contents

PREFACE—*XI* ACKNOWLEDGMENTS—*XII* INTRODUCTION—*1*

NARRATIVE
- I The Residence Act—*2*
- II The Quarry on Aquia Creek—*4*
- III Highlander, Stonemason, Overseer—*6*
- IV A Powerful Base—*8*
- V Edinburgh—*11*
- VI Workmanship—*14*
- VII A Village of Workers—*16*
- VIII From Quarry to Stone Yard—*18*
- IX The Finished Carving—*26*
- X Stonemasons, Brick Masons, and Apprentices—*32*
- XI President Washington's Last Visit—*35*
- XII The Great White House—*37*
- XIII The Walls Rebuilt—*39*
- XIV Adding the Porticoes—*41*
- XV President Truman Saves the Walls—*43*
- XVI Preservation of the Stone—*46*

NOTES—*48*

CATALOGUE
- I Aquia—*52*
- II Enter James Hoban—*76*
- III It Is Agreed—*82*
- IV The Finest Stonework in America—*86*
- V Stone for the Porticoes—*100*
- VI Survival and Preservation—*122*
- VII Stone in the Monumental City—*134*

REFLECTIONS—*148* ILLUSTRATION CREDITS—*150* CONTRIBUTORS—*152*

Preface

LAST FALL WHEN **William Seale and I walked among** the rocky outcroppings at the old Aquia quarry, we could almost hear the echoes of the workers there cutting stone and moving it along to the river to be rafted up to Washington to be used in building the young nation's new President's House and Capitol. That great stone quarry is testimony to their labor, two-hundred years ago, in the now quiet woods of Virginia.

Ten generations ago my ancestors left Scotland and entered America through the Cape Fear Valley of North Carolina. But there were others who came from Scotland in the eighteenth century, whose skill and craftsmanship were sought after for being the best in the world. These were the stonemasons who brought their expertise to the creation of the stately walls of the President's House, the image of which has been symbolic of America to billions of people around the world for 217 years. Not all the workers who built this great house came with renowned stone carving talent. Many were laborers, enslaved and free, whose sweat and muscle gave it shape.

Through the fires set by the British in 1814 and the major renovation of 1902, the exterior stone walls created by the Scots stonemasons remained. Then, in 1948, when it became necessary for President Harry S. Truman to remove the entire interior of the White House and reframe a solid infrastructure to make the house sound and safe, he demanded the original walls be preserved authenticating the White House symbol for all time.

This book is about the stone carved from Virginia rock, transported along the Potomac, shaped on the riverbanks in Washington, moved to a stone yard for final carving and positioning, and put in place on a mansion for the president. In a sense those who labored are the stone foundation of our country as well, and future freedoms would be realized and signed into law in the very building that their hands worked to construct.

Stewart D. McLaurin
President, White House Historical Association

Acknowledgments

MY CONTINUING THANKS TO THE LATE Ierne Grant of Edinburgh, Scotland, who offered guidance in research and first discovered the White House stone masons in the records of the Masonic Order in Edinburgh. I am grateful as well to the Grand Lodge of Edinburgh, oldest of all Masonic lodges in the world and founded, of course, by stonemasons, for its assistance and the opening of its archives. For help with research at Aquia Quarry, I thank Jane Henderson Connor, preservationist.

At the White House Historical Association in Washington, D.C., Stewart D. McLaurin, president, showed great interest in the subject and suggested it be revisited in this new book. Marcia M. Anderson, vice president over publications at the Association, first called the idea a book, made it part of her list, and supervised every aspect of its production, while listening to the author interminably. Ann Grogg, consulting editor, shared her high editorial standards and knowledge of history, cheerfully rescuing me at times from possible mortification. Kristin Skinner, production manager, I have come to see with armloads of papers and a keen eye on the screen, the able keeper and organizer of details for this book. Many others at the Association were involved in different ways, and I wish to thank Fiona Griffin, Lauren Zook, and Olivia Sledzik in the Publications Department. My thanks go to photographers Martin Radigan, who documented the beauty of both the Aquia and Seneca quarries for us, and Bruce White, who once again captured the fine stone detailing of the White House walls, and to artist Dahl Taylor for his original illustrations of the stone carvers at work.

The visual record would not be complete without the assistance of those at the White House who graciously accommodated our photographers, most notably Chief Usher Angella Reid, Curator William Allman, and Assistant Curator Lydia Tederick.

I am indebted to scholars who have searched this general subject before me. Harley J. McKee, Lee H. Nelson, and, more recently, Robert Kapsch shadow every page of this book with their own outstanding research and discoveries on the stonemason's craft and the workers employed by the government to build the White House. I thank Gary Thomas Scott for being a ready resource on the Masonic Order. Paul Dolinsky, chief of the venerable Historic American Buildings Survey, and James Jacobs, who carried out an amazing study of the White House while the old paint was being removed and created a comprehensive set of ink-on-linen drawings of the building, more fine art than mere recordings. Stored in the Library of Congress, these drawings are one of the great achievements in the study of historic architecture in our generation. They have been used extensively in writing this book.

William Seale

INTRODUCTION

The Stone Beneath the Paint

O N AUGUST 2, 1792, George Washington, president of the United States, drove a stake in the ground locating the residence of the president. It would be the first official building completed in the new capital that was already named for him. Washington's stake replaced another nearby, the forgotten stake driven a year before by his now-disgraced city designer, Pierre Charles L'Enfant. The French engineer had located on the same spot a "palace" of his design built of stone. Washington retained the idea of stone, but built upon his stake a smaller building for the same purpose, a house for the president that history would call the White House.

This book is about the stonemasons and stones of George Washington's White House. It is drawn from documents public and private, governmental and institutional, American and Scottish.[1]

Sources here are not restricted to the written word. The stones in the walls of the President's House are not silent if you look beneath the paint that makes them white and focus your attention on their placement, marks, scars, floral carvings, and varied shapes. Down the Potomac River, the Aquia quarry from which the stones came remains. Abandoned and apparently left as it was two centuries ago, it yields volumes to the eye about how hundreds of men transformed the raw outcroppings into building blocks.

The stone walls are all that is left of the original White House. They remember great moments and the full succession of presidents.

CHAPTER I: A NEW FEDERAL CITY

The Residence Act

THE UNITED STATES Constitution gave the United States Congress "exclusive Legislation" over a yet to be established district that would be the "Seat of Government." Thus Congress, seated temporarily in New York, passed the Residence Act of 1790, establishing a new, permanent U.S. capital, to be occupied ten years hence. It would be a federal territory, 10 miles square, flanking the Potomac River, on lands ceded by Maryland and Virginia, near the middle of the new union of states. The act further provided for "suitable buildings for the accommodation of Congress, and of the President, and for the public offices of the government of the United States."[2]

Except for the river ports of Georgetown (Maryland) and Alexandria (Virginia), the new Federal District consisted of farmland and woods. The site of the Federal City was to be purchased from landowners. What was soon called "Washington" would stand at the highest point a ship could sail inland from the Atlantic, more than 160 nautical miles from the mouth of the Chesapeake Bay and up the Potomac to the fall line. George Washington believed that future improvements on the river would connect it to the Ohio River Valley and thus westward to the great Mississippi and the Gulf of Mexico. In his mind, the new American capital had every opportunity to become a new Paris, the commercial and political center of the young republic. Washington wanted his city to realize for all time the symbolic identity its importance merited. There being no interior department, normally federal building projects fell under the secretary of state. In this case the president pulled rank, assuming the final authority himself. Thomas Jefferson said, with apparent relief, that the president would be too busy to supervise details.[3] The secretary of state was mistaken.

Washington's interest was in his urgency to occupy and thus confirm the new capital. Ten years was too short a span to even imagine completing the entire objective specified in the Residence Act. Thus he turned special attention to what seemed the most attainable goal, the house for the president. He knew politics: without a finished public building in place, and thus little to show for the money spent, Congress, now meeting in Philadelphia, might modify the Residence Act and move the capital someplace else. Rival presidential mansions were under construction in New York and Philadelphia. As the 1790s passed along, Washington, although coaxed, would enter neither.

Washington wanted the president's residence in the new Federal City to be of stone to compare favorably with those of world capitals over the ocean. Paperwork on the proposed house mounted in quantity, enough of which remains to show that Washington's vision of an appropriate style was rooted first in L'Enfant's idea of a palatial house about four times the size of the White House to come. The footprint was described precisely in L'Enfant's plan, and excavation of the cellar had begun, cut deep in the reddish soil. In February 1792 L'Enfant departed, following quarrels with the city commissioners appointed to oversee the building of the Federal City, and eventually Washington became dubious about him. Fearing delays caused by high cost and the difficulty of obtaining materials, Washington permitted considerable reduction in the size of the proposed house, at the same time calling for elaborate stone carving.

The next month, in March 1792, a competition was advertised for the design of the President's House, and it was won by an Irish immigrant architect and builder named James Hoban. Friends in Charleston, South Carolina, had presented Hoban to Washington while the president was on his southern tour in the spring of 1791, well before the competition was even announced. Washington was almost certainly shown some of Hoban's work, and whatever he saw impressed him. As the competition proceeded, the president personally saw to it that Hoban was invited to enter. Hoban hastened to Philadelphia and gave his ideas to the president. He went then to Georgetown, where the commissioners met, and made his competition drawings for the President's House under their eyes and those of Washington himself.

The amiable Irishman was by no means the amateur that his later nemesis, the architect B. Henry Latrobe, jealously called him. A peasant boy from County Kilkenny, Hoban had strong connections to his parents' employers, the powerful, landed Cuffes, Cromwellian lords of Desart Court, the great estate where he had been brought up. Probably through their influence, perhaps also that of his Hoban cousins, quarriers of the famous black Kilkenny marble, he had been admitted as an apprentice in the Dublin drawing school of the distinguished Anglo-Irish architect Thomas Ivory, on a scholarship from the Royal Dublin Society. Hoban was privileged to work under Ivory and later with Ivory on several of Dublin's major public and private buildings, including the mercantile Royal Exchange.[4]

Remembering his Irish experiences, and sizing up his presidential client, a wise Hoban sweetened his proposed President's House by modeling it after the finest house in Dublin, not a public building but the private mansion of the powerful Duke of Leinster, an Anglo-Irish nobleman universally admired as the "First Gentleman of Ireland" who was both the sponsor of the Dublin school and the chair of the Royal Dublin Society. To Washington the parallel of president to first gentleman was perfect. Hoban turned the idea of Leinster House into a competition design. Washington and Hoban seem to have made many modifications, dropping one story and greatly simplifying the interior, but the original architectural vernacular of the mid-Georgian Dublin model was not lost.

Washington was pleased with the design and the tone of the architecture. He was also pleased with James Hoban, observing his energetic manner, his desire to please. And he was right in his choice: the house would stand complete eight years later, when the government moved to the Federal City.

CHAPTER II: THE PRESIDENT SPECIFIED STONE

The Quarry on Aquia Creek

THE COMMISSIONERS HAD begun a year earlier, in 1791, to prepare for construction as they knew best. In their minutes and papers, they seem frantic, running in many directions, fearing the president's censure. Even so, they accomplished many obvious objectives. Bricks, timber, and hundreds of iron tools were assembled at the site of the President's House, as readily as they might have been for building a church or courthouse, only in greater number. Constructing the underlying packed-brick rubble foundation upon which the basement walls of stone were to be built was familiar territory for the brick mason's art had been well known in America for more than a century. Clay pits were authorized, built a short walk from the site. Lime burners were engaged to open their seasoning pits. Laborers and skilled workers, both white and black, accomplished the preliminary work.

Stonemasonry was a greater challenge. Hoban was a skilled engineer of stone construction and designed stonework well, but he was not a mason himself. When the president specified stone, he did not mean fieldstone, rock, or the blocks typical of his relatives' rough cut-stone houses in Virginia's Blue Ridge and buildings particularly in the German settlements of other states. He wanted refined, dressed stone walls, and rich carving.

Meanwhile, the matter of obtaining stone for the Federal City's buildings had already begun. Brent's quarry on Aquia Creek, 45 miles downriver, in Stafford County, seemed the obvious source. Aquia Creek was navigable, running a few miles inland from the Potomac. The actual quarry was on the 17 acre Wiggington's Island. It had been harvested for a hundred years on a small scale, for tombstones and doorsteps; now it was to be quarried for monumental buildings yet to be designed. Probably the largest Aquia stone construction was the west facade of the Carlyle House (later replaced with limestone) in Alexandria, Virginia. Aquia stone produced door and window surrounds at colonial Church of England buildings in northern Virginia, standing today as Episcopal churches. The wall enclosing the Masonic Cemetery in Fredericksburg, Virginia, is of Aquia sandstone, as are Mount Vernon's pavements and doorsteps. Less dense sandstones, like Aquia, were easily cut materials called for that reason "freestone."

Freestone soaked up water at varying degrees depending upon the stone's general character, but that from Aquia drew enough to make one worry. Had the builders been in less of a hurry, using this stone for great buildings might have made them pause. But they were in a hurry. The stone was often rubbed vigorously and whitewashed in an attempt to arrest the water problem, a technique not wholly different at that time from sealing the clay walls of cellars and graves.

The Aquia sandstone quarry was accessible, and haste was the motive, so the government purchased Wiggington's Island in 1791 and contracted with carpenters to build workmen's huts and a kitchen. The first twenty-five quarrymen were African slaves hired from their masters. They began pulling and cutting away the tangled woods that covered the giant swells of sandstone. Contracts with the slaves' masters were specific, assuring that each workman was well fed, well housed, properly medicated when necessary, and well clothed. The masters of the hired slaves each received a fee of from about $130 to $150 a year.[5] Soon the crew expanded in number, remaining from then on about forty to sixty men, both white and black, assembled under the direction of Captain Elisha Williams, a Scot with experience managing quarries. He directed his men in preparing the island for a much larger stone operation than the Brents had known. Once the vegetation was gone, the workmen hacked and split the smooth, boulder-like surfaces into smaller, more manageable blocks. By Christmas 1792 the quarry was a village of workers.

Despite concerns that the quarry with such demands upon it might exhaust over the years of building Washington, Aquia continued its yield. Sounding the great outcroppings with poles of iron or wood and guesses by experienced quarrymen were at that time the only ways to know how deep a stone deposit might be, so an experienced ear, speculation, and strong faith kept the Aquia quarry going. The commissioners of the Federal City leased adjacent properties, extending their possibilities for stone, yet Wiggington's Island seems to have been the main source.

CHAPTER III: COLLEN WILLIAMSON

Highlander, Stonemason, Overseer

THE NEXT PROBLEM for the commissioners was finding stonemasons qualified to build monumental architecture. America was not a total void for large-scale stonework, particularly in the city of Philadelphia and, to a lesser extent, New York. James Traquair, a stonemason and carver originally from Scotland, was a very busy and highly thought of contractor in Philadelphia, owner of Wilkinson's quarry of limestone and "colored marble," one of ten in the Philadelphia area. Jefferson inquired of him for the commissioners, and Traquair was as helpful as he could be, knowing that there were building projects in the Federal District coming up.

It was probably Traquair who made the stonemason Collen Williamson aware of the President's House project. But Williamson, it turned out, had closer connections than Traquair could have provided. The building commissioners met in Suter's Fountain Inn in Georgetown, a prominent hostelry run by John Suter, Williamson's cousin from Scotland. This placed Williamson directly in the company of the commissioners, who seem to have been so happy to meet a bona fide stonemason that they bound him by agreement as a sort of general supervisor of stonework on the President's House, with a mandate to assemble twenty or thirty more masons.

We know more about Williamson's earlier life than we do of any other stonemason on the project. The big talker and self-promoter was born in the sandstone region of Grampian, Scotland. Baptized in the little stone church in the village of Dyke on June 23, 1727, he was 65 years old when he first stood before the commissioners. Son of a stonemason, he had followed the trade all his life, apparently remaining in his home region. He developed great skill, and while in his late 20s or early 30s began working for Sir Ludovick Grant, of Castle Grant, the old seat of Clan Grant, great landholders. For a local mason to achieve favoritism from so rich and powerful a man was reason enough to keep him a country builder, rather than gravitate to a city that offered more work. His stonemasonry usually followed the designs of the Adam brothers—John, James, and Robert—natives of Scotland, who interacted often with the Grants and their projects.

In about 1758, Sir Ludovick decided that Castle Grant, which he had been remodeling with John Adam, was too big and drafty. Sir Ludovick was a widower with older sons elsewhere and unmarried daughters still at home. Williamson was asked to draw a house for the laird at Moy, where he planned to pull down an existing house, build a new one, and go there to live most of the time with his family. The stonemason obliged and began construction on a large, rather typical, castle-like Scottish gentry house, very vertical, with the impression of massed square towers.

The work went well at Moy. Williamson had the walls built by 1764 and was at work on the interior. A domestic staff had already been assembled, although the house was not yet occupied. Suddenly Sir Ludovick's authority over the estate was challenged successfully by his eldest son John, living in London, claiming that the laird was mentally incompetent to manage the estate. The actual charge was drunkenness. All work stopped at Moy. Builders and staff departed. When the work resumed, it was turned over to John Adam to complete.[6]

Where Williamson went from Moy is unknown. He appeared at Suter's Fountain Inn in 1792. Before that time he insisted he had worked in New York, but no details are found. One thing seems clear: he had cast his lot with the Grants and in middle age had been cast aside. He never forgot his aristocratic association in Scotland, dropping Sir Ludovick's name and that of John Adam with some frequency, justifying himself as an intimate of his former high connections. Moy House stood for more than 250 years, expanded and improved, only to burn to Williamson's stone walls in 1995. Today it is a ruin enveloped in its old garden, whetting the appetites of real estate developers.

At the time of Williamson's agreement, in the summer of 1792, Hoban was working in the commissioners' office in Georgetown, completing his competition drawing for the house, which was accepted by President Washington and the commissioners in October. The other entries, including one by Jefferson that must have amused Washington in its oddity, were rejected. Hoban's house was final. There would be very few changes to come. Williamson's contract required him to direct removal of stone from the quarry to a stone yard near the building site. He was then to start the work cutting and laying the stone on the brick foundation of the President's House.[7] He went to work, doubtless understanding the innocence of the commissioners about building. The surviving documents reveal excitement on his part, and the gratification of the commissioners that he had come along. Williamson was also to design and build a house for the commissioners in the district. He submitted his design June 7, 1792, for a large house with an elegant stone doorway and eight rooms with fireplaces. The planned house, which apparently was never built, is typical of the spotty demands of the commissioners and Williamson's ultimate frustration with them.

CHAPTER IV: THE FOUNDATION
A Powerful Base

WILLIAMSON'S WORKMEN NEVER numbered near the projected thirty; at most, they numbered eighteen. Four or five stonemasons may have worked for him before in New York. They were paid by him, and thus their names are lost. Others were African Americans, again with names that are lost if they were slaves, because their masters received the wages for them. Or some payments went to middlemen whose business was "negro hire."[8] Free men's names seem to have been listed in full, but their racial identity was not recorded and cannot be known.

Williamson's work was superb, but his temperament was terrible. His ethnic contempt centered in the Irish. How he loathed them, describing James Hoban's Irish workmen as "vagabons." Soon enough his dislike carried over to Hoban himself. Williamson claimed that he was as much an "archeck" as Hoban, citing Moy House.[9] He resented Hoban and what Williamson insisted was his inferiority to hold general authority over the works.

Williamson and his lodge completed the raised basement of the President's House with notable speed. He created the first and lowest foundation by digging down, a 4 foot deep ditch like a trench all around the outline of the house, filling it with gravel and rubble and tamping it hard, thus making the base for the 12 foot basement walls yet to rise, a vertical total of 16 feet. Recall that the foundation and basement of the house were built inside the large excavation intended for the earlier plan and would need extensive filling around the edges to adjust to Hoban's plan. The basement stone walls were meant to be exposed and therefore handsomely finished on the outside and smoothed with less finish on the interior. They were nearly 4 feet thick, solid stone. Exterior surfaces of the stone walls were flat, made up of ashlar or uniform rectangular blocks. Sizes of the ashlar blocks did vary. The outside face of the bow upon which the South Portico would one day stand was decoratively formed to feature large protruding stone blocks from the otherwise flat surface, with sufficient space between the "blocks" to enhance their appearance of massiveness. This architectural base was rich and heavy, visually boasting its support of the large half circle of the walls that would be built upon and above it. The device was reflected in framing "keystone" blocks alternatively around the basement windows.

From the bottom of the long slope to the south, the rising building seemed too small for the long east-west ridge upon which it was built. It still seems too small, even though it is visually corrected with plantings today. From the north it does not suffer this problem. Hoban turned the sharp differentiation of grade to his advantage, and now that the basement was built, the wisdom of his solution was clear to all: the bulk of the building he slid about 40 feet into the ridge from the south to north, leaving the south facade, including the basement, fully exposed, while on the north the basement was below grade and not visible. On the south the sun dried the masonry walls of the house and warmed and lighted its rooms. A sunken areaway on the north, out of view, pulled the earth away from the basement level there and made a broad air space or paved areaway to allow light to come into the basement and to provide unseen carriage access to the kitchen and other service rooms of the utilitarian level of the house. Although the basement on the north did not share the warming sun, it was to contain the 40 foot kitchen with its two great fireplaces on opposite ends. The rest of the basement rooms were storage spaces on the north and servants' rooms on the south.

The upper two-story walls were not as heavy as the stone walls of the basement. Walls here were backed with a 2½ foot inner lining of brick. For the President's House the brick was an oversize, soft, clayish brick molded and fired in igloo-type kilns on the site, and baked low-fire because the resulting bricks would never be exposed to damaging weather. In density the White House bricks were only somewhat stronger than adobe. The stone was the face and armor of the building, but in total the maximum 14 inch depth of the stones was extended by the brick backing to walls nearly 4 feet thick. This technique of brick-backed stone construction was familiar in the eighteenth-century building world of Williamson and Hoban and long before that. Solid stone building, except for small houses and stores, was too costly, and technically solid stone construction, without air space between the vertical layers, threatened future deterioration from trapped moisture. At the White House, the usual barrier of slate, laid flat, created a moisture barrier between the solid stone basement walls and the stone and brick walls above them.

Williamson's contempt for Hoban rose to storms of anger. One might speculate that the quarter-century difference between their ages had some influence. The more politically astute Hoban, aware that the outbursts only annoyed their superiors, kept quiet. It bothered him not at all that his Irish boys drank and played after work hours, if their labors for him were satisfactory. He delivered no harsh reprimands, but disciplined them in his own way, with a system of fines. He required their membership in the militia company he captained. When a worker (all workers were male) skipped muster, he was charged a fine, which Hoban subtracted from his wages. There was no recourse. A misstep on the job, such as a fight or harassment between workers, or drunkenness, also earned fines. Why Williamson, who reported this to the commissioners, would think that this arrangement was any of his business must have puzzled Hoban, for it worked effectively.

At last, at the close of the building season in 1795, the commissioners declined to renew Williamson's agreement with them. He had cut off his nose to spite his face. The commissioners did not like dealing with him, vastly preferring Hoban, whom they came to treat rather like a son. Perhaps Williamson's ravings convinced them that he lacked the stability to build the big house for the president. Whatever the case, they had already sought out new stonemasons for the project.

Williamson's dismissal was not pleasant. Some of the money due him was withheld. He spent the rest of his life pursuing fair payment, which never came. Years later, he wrote to the commissioners in support of his application: "I served under the tuition of one of the first Archetects in the world, the great John Adam" and "I understand Archectry equele to the most of men. . . . You was told by one of the first gentlemen on the content that I was a man of varasety and master of my business." Public building commissions are characteristically heartless, and these, whom he called "poor blind commissioners," were no different, in his estimation.[10] Until his death in 1802, Williamson otherwise fared well. Now and then he participated as the expert in technical procedures and was even invited by Hoban to help out. He and his wife Margaret lived in town, in a house they owned, comfortably furnished, and they profited from rental properties. In 1797 he said that he had built a brick Presbyterian church in Baltimore, but he would live only a few years more. His inventory following his death in 1802 includes accoutrements of a small farm but not a single implement to suggest he continued his stonework.[11]

But men of Williamson's skill were sorely needed to finish the house. The commissioners wrote to American agents in Paris and elsewhere in France, only to have their letters returned with a warning that the state of the French Revolution was such that any request for Frenchmen to leave France might touch off real trouble. With news of the Reign of Terror crossing the ocean to America, the agent did not have to amplify what he meant. The commissioners sent letters elsewhere in Europe, with no luck. They returned to James Traquair in Philadelphia for references, and he produced a list of stonemasons he had known in Edinburgh. Several inquiries were sent, but no responses came back.

George Walker, a well-known Philadelphia merchant very much invested in lots in the new capital, an unsmiling but effective businessman who closely watched activities in the Federal City, was leaving for his native Scotland, via London, on a buying trip for his store. He agreed to take a letter from the commissioners to Edinburgh, to seek stonemasons. This he did, and the story of the stonemasons of the White House took an entirely fresh turn.

CHAPTER V: CONSIDERED THE BEST
Edinburgh

SCOTS STONEMASONS WERE favored for major projects all over Europe. Some of the stonework Jefferson praised in France was likely the work of Scots. The craftsmen had been willing to travel as a means of escaping economic uncertainty in their own country. Scotland, joined with England following the Act of Union of 1707, was a country of stone buildings. Long held back in its economy by the wars and uprisings of the Jacobite rebels, it had slowly begun to flourish in the late eighteenth century. Glasgow's linen mills rolled and expanded. Edinburgh threw down its containing walls figuratively and, it will be seen, literally, becoming a modern capital politically, educationally, and culturally.

The great skill of Scotland's stonemasons showed in superb architectural cutting, finishing, and setting, always considered the best, and as stone carvers they were rivaled only by the florid delicacy in the marble work of the Italians. Scots worked in Britain, of course, and in France, Italy, and Spain, and in Russia they built palaces for the czars, notably a black marble bathing pavilion for Catherine the Great at Tsarskoye Selo. The finest architects and patrons of the time sought their skills and brought them to buildings public and private alike. The greatest concentration of the best stonemasons in Scotland was in Edinburgh. There in the 1780s and early 1790s they enjoyed a field day of building. Central to the several decades of boom was the development called New Town. Built on land formerly considered useless, and feared for crime among its residents, New Town was to replace with new and more tax-worthy construction the old shanty slums that rambled along narrow streets, over the steep, stepped ridges that fell like ruffles on a fancy cuff from the old medieval city, high above, that had once defined Edinburgh.

When the city further proposed an upscale residential transformation to a rather dark part of New Town, the magistrates wanted a development of streets and garden-like squares. The Adam brothers were engaged to create the master plan. Their plan, celebrated today as a monument in town planning, called for demolition of all that was there and garlanding the site with beguiling sweeps of programmed architecture flavored with their restrained, decorative style of neoclassicism. The business concept for the Adam plan was to be realized through private development. A covenant in the deeds of the lots required the investor to build the facade of the buildings of stone following the Adam brothers' design. Behind this prescribed stone facade, the owner could build anything he wanted.

The plan presented a dream opportunity to men in the building trades. Stonemasons had a ready market for building facades, and brick masons, carpenters, and glaziers for the rest. For more than a year, New Town's fresh-laid streets were alive with construction. Huge quantities of building supplies, hauled in oxcarts, awaited the skilled hands of master builders and crews of craftsmen, laborers, and apprentices. Stonemasons were in first demand, thanks to the facade requirement. The project promised new life to Edinburgh. Promoters, tradesmen, and realtor speculators saw wealth pour heavily on the city. Lady Margaret Grant, one of Sir Ludovick's kin, toured Edinburgh and wrote to Lady Pitcalnie, "How wonderfully the building goes on, you will scarce find your way in this town."[12]

The ultimate success of New Town speaks for itself, but in autumn 1793, when George Walker reached Edinburgh, he found New Town an empty ghost town. The Adam brothers' neoclassical fronts that stood were unfinished, silent, dark, with vacant sockets where windows and doors were intended. Building materials were stacked, unused, guarded by hired watchmen with dogs. The cause was the war with Revolutionary France, which had begun the February before. In swift reaction to French aggression, Britain had made emergency preparation including the resurrection of acts of 1750 and 1782 that prohibited construction or the emigration of skilled labor. They had been enacted originally to stave off emigration to the financial attractions of Europe and the New World. Now the motive was to keep the laborers at home, standing by during the French war. New Town development was frozen in its tracks. Only banks and merciless moneylenders were still about with their mortgage papers. Many skilled tradesmen had themselves invested in a New Town lot or two, with the idea of erecting a house and selling it for a gratifying profit. Weighted by debt, the building men were like those in a shipwreck, most aboard knowing they were doomed but the inevitable stronghearted not yielding hope.

As instructed by Traquair, Walker made his way to Shakespeare Square to find Alexander Crawford, "a warm frind to America," wrote Traquair, and a man thinking of moving there after his work in New Town.[13] Alas, Crawford had died. His widow Margaret Crawford sent Walker to the master's partners, James and John Williamson. Circumstances being what they were, with heavy mortgages pending, John Williamson agreed to sail to America. Walker sought out other stonemasons at the Grand Lodge of Edinburgh, the oldest Masonic institution in the world and before 1600 allowing admission only to stonemasons. He was directed to Lodge Number 8, an operative Masonic chapter (meaning restricted to working stone men) to seek further stonemasons. It was a new, very exclusive lodge, a "breakaway" subsidiary of 1788 that existed within the larger envelope of the Edinburgh Masonic Lodge and preserved the strict tradition of a limited master stonemason membership.

George Walker was taken in as a friend and readily signed six more masons for the White House, all leaders in their craft: Alexander Scott, George Thompson, James White, and Alexander Wilson, plus two fine carvers, Robert Brown and James McIntosh. Their signatures are in the Masons' membership book. Inscribed beside White's name was notice that his dues would be resumed "When he is again in the country." Another signature, that of Alexander Reid, had inscribed by his name "in America." Today in the Masonic Lodge Archives in Edinburgh the small, leather-bound tally of membership that records these signatures survives and, indeed, is kept current these 250 years later.[14]

The Edinburgh masons sailed from the west shore of Scotland to the Virginia port of Norfolk in America. Each man knew he was breaking the law by leaving Scotland, so their financial urgencies must surely have outweighed any hesitation they had. Their objective was to make money to send home, to rescue their failing businesses and save their assets in New Town. They left secretly, in the flow of people emigrating for many reasons, few altruistic. Most of the stonemasons would return to Scotland. A few would remain in America.

CHAPTER VI: WORK AND PAY

Workmanship

THE COMING OF the Edinburgh Scots to the building of the White House in 1794 marked the appearance in a rising Washington, D.C., of modern European tradesmen, top men in their field, which was at the top of any building project. Those who were called master masons knew no men higher than they in the actual work of construction, and in the case of Scots, each man was skilled in every phase of stonework, from cutting to finish. Surprises awaited them. Not only was the landscape new, but their treatment as what would be called today professionals was not what they had known from clients back home. James Hoban gave them no difficulty and, having worked with Scots in the building trade in Ireland, probably understood the difficulty a European stonemason might find dealing with these commissioners of the Federal District, who would surely claim unreasonable authority over them and, for all the trouble of securing their services, would consider them no more than common laborers because they worked with their hands. Many Americans shared this view, even though most Americans' hands were no cleaner. Any deference the Scots stonemasons might have accumulated over the ages in Europe with their hard-earned skill was of no interest to the commissioners, only its application in building the house for the president.

One early and outstanding conflict between the two was the manner of payment. The commissioners ordered Hoban to arrange payment on a basis of daily wages, a flat rate to which some benefits were attached, often food and housing. To the stonemasons wages were demeaning, casting them into the unskilled laboring class. They took pride in their work. They were accustomed to being paid after it was done, when they could present it for "measurement" and proudly claim fair pay. The ancient formulas of measurement were familiar to all building people, but especially to Europeans, where they had been current at least since medieval times.

Determining payment by measurement had nothing to do with time or labor, but with the quality of the work done. It might be applied to an entire building, or, say, a 20 foot section of a wall. Measurement worked this way: at the commencement of the work, tradesman and client mutually selected a "measurer" as their expert judge. Usually a respected senior in the trade, the measurer was called in when the work agreed upon was considered by the workman to be complete. The measurer applied one of many formulas to the product submitted—

volume, scale, workmanship—in coming up with a price, to be in the range of what the client and workman had agreed upon. If the client or workman disagreed with the measurer, negotiations would follow until an agreement was made for final payment. If this proved impossible, the matter was taken to court. Records of such disputes are numerous in European and American court records before the Industrial Revolution and the proliferation of shortcuts.

The commissioners, with their eye to precise bookkeeping, believed that wages provided more accurate records of work and had already established wage payment on the job for other workers. The arrangement with Collen Williamson had settled down largely to wages, an average of which was about $30 per month, apparently with occasional measurements with extra fees. That his crew was not called on to carve the stone, however, gave the Edinburgh men an advantage in higher payment after he was gone, and they held out. Throughout the project, a continuing bulldog tug-of-war kept the masons and the commissioners pressing for a solution. Not a whisper of an agreement was recorded. The building went on. No compromise is confirmed in the written documents, but more than 150 years later an exposure of marks hidden inside the structure would reveal the answer. While wages did continue with some, most of the Scots had their way with measurement.

Most of the issues with the commissioners were handled by Hoban without much difficulty for the stonemasons. In an effort to save money, the commissioners every year, at the beginning of the work season, attempted to reduce wages, seeming to think they were in the driver's seat and that the workers had no place to go if they left over the reduction. As time passed, however, and the deadline became an ever greater challenge, tables turned. The commissioners feared losing their skilled workmen and assumed both a milder manner and opened a more generous purse. In truth the commissioners knew very little about building and often faked more knowledge than they had. Dr. William Thornton, architect of the Capitol, was the masons' best friend on the commission, if not for any sophisticated knowledge of construction he might provide, then certainly for his intense interest in the subject. The brilliant, scrappy, gambling doctor-architect seems to have considered himself one of them and fought many a battle for all the workmen, not only for the stonemasons.

With the stonemasons at work, the building of the President's House could now move forward with little interruption toward completion. For George Washington it was a more important work than even the Capitol, itself in a continual tangle of architectural controversy. Creating symbolic buildings as these two were was a unique exercise. It was a confusing and complex challenge to dissect and work out, requiring philosophical thinking as well as mortar and stone. The Capitol was to serve hundreds of people, a Congress, clerks, and the visiting public. It would also accommodate the Supreme Court. By contrast, the presidency, an equal branch of the three components of the federal government, required an architectural statement of a president's presence that could be satisfied with a house. An elected president, one man, would fit very nicely into a large house, a fine residence where all his official work also took place. Finishing a house, even a big one, turnkey, could be more easily relied upon than an ambitious though greatly unfinished Capitol to establish Washington forever as the seat of government for the new nation.

CHAPTER VII: A TEMPORARY SCENE

A Village of Workers

JOHN WILLIAMSON, one of the Edinburgh stonemasons, was at times the master of his part of the project; sometimes Alexander Reid took the role, which appears to have been passed around. The workmen were provided with small wooden cottages, one room about 12 by 12 feet, the little houses like the others lined up soldier-like to suggest a village. They were crowded, the domiciles of many. With the stonemasons in the village were carpenters, bricklayers, and many trades that came in for short times, such as the shell-lime burners from Frederick, Maryland, who drew from their earth-seasoned pits for this public work. It seems clear that the Edinburgh men privately kept well apart from the others.

The rough, ungraded land around the building was stacked with materials. It was not pretty, but the absence of crime and convenience to work made it a good place in which to live for a time. It was as temporary a scene as might be imagined, for when the President's House was complete, all of it would disappear. The men who built it were a diverse group, from the politically harangued gentlemen commissioners, to the indispensable Hoban, to the men in the various trades, to the laborers, some of whom were slaves. The lines of small, square unpainted frame cottages grew numerous. Others were built where the West Wing stands today. Marriages and immigration of families eventually filled some of the houses with wives and children. Hoban pulled two cottages together when he married, providing an enviable domicile for the head man in the new city of very few houses.

An extensive brick yard, built to the north of the house, was in full operation by the spring of 1793. Other work areas appeared as close to the village as possible. Enclosures for animals, oxen, a few horses, and some mules were on the west. Storage sheds, a vast stone yard, and both open and roofed stonework sheds were crowded together on the east. A distance from the building site, on the north, the commissioners built a hospital with a resident nurse and a succession of four doctors on contract over seven years. The workers received good care.

In what was to be Lafayette Park, where Andrew Jackson's statue stands today, a neatly finished wooden shed, only partially enclosed, with wooden floor and board roof, was carpenters' hall, a large work space 50 by 25 feet, built by William Knowles for the commissioners. In this sturdy shelter on workdays the carpenters chiseled and sawed their mortise-and-tenon floor joists, planed rafters, and shaped with the adze the great beams that would hold the broad spaces of the house in their embrace. On Sundays, to this simple shed the Scots brought Presbyterian worship, founding a church they called St. Andrew's. Baptists, Catholics, and Episcopalians worshipped there in turn. Father Anthony Caffrey, wanting a permanent building for his flock, persuaded Hoban, a devout Irish Roman Catholic, to build St. Peter's Church a few blocks away. Also in carpenters' hall the Scots and fellow workmen, including Hoban, Collen Williamson, and some commissioners met as Lodge Number 15, the enduring Masonic Lodge still thriving in Washington as Federal Lodge Number 1. Hoban and Williamson were both active in its creation, Hoban being grand master. By the late 1790s he was known to all as Captain Hoban, for his militia company, the Washington Artillery, and in all the chaos of the building site he mustered them out for drills on Saturday afternoons.

Saturdays were also market days. Workmen planted vegetables for their own tables and also to sell. A market developed, where vegetables, fruits, meats, country wine, and hard spirits were sold, as well as the workmen's produce from their garden plots. Saturdays became festive occasions, with dancing, games, and horse racing.

The construction scene as the walls rose must have reminded one of Hendrick Avercamp's sixteenth-century Dutch genre paintings, showing crowds of individuals and groups at their daily pursuits. At the doorways of cottages, wives cooked and sold food, sewed and washed, tended chickens, and drove foraging hogs away, while the men labored upon the building, tools in hand, looking down ever higher from ladder, scaffold, and windowsill. The higher a man climbed in working on the walls, the greater his whiskey or rum allotment for the day. The commissioners' records in the National Archives paint a word picture of this scene over many documentary pages.

CHAPTER VIII: CAREFUL PLANNING

From Quarry to Stone Yard

THE PROCESS OF building the walls of the White House began at the quarry. John Williamson or the current master of the stonemasons, often with Hoban, sailed or simply floated downriver in the current to the mouth of Aquia Creek, then traveled up the creek to the island, where they met the quarry master in the sprawling stone camp. There laborers trained for the work to split the stone from the greater rock formations by one of several methods. Aquia stone is a relatively porous sandstone, technically known today as arkose sandstone from the Lower Cretaceous period. Harvesting it was simpler than that for the harder limestone, but its imperfections were sometimes more.

The usual method of extraction from the Aquia outcroppings, as evidenced by surviving marks today on the stone, was to drive a series of carefully whittled round green sticks 4 or so feet into a natural cleft or drilled hole in the stone and to pour water slowly and generously into the exposed end of the stick, which causes the capillaries in the stick to swell and thus split the stone. After the initial split, if the crevice was not complete, the long-handled wrought-iron lever called a "jack" was inserted to pry the split to completion. A skilled hand could achieve a very significant, quick split, like a slice of bread falling from its loaf.

Splitting was the simplest means and, by surviving physical evidence, was the one usually employed at Aquia. More difficult to master was the jumper, a forged-iron rod, pointed at one end, flat-headed at the other. This process involved several men, one to penetrate the rod in the stone some inches, then hold it steady so two or three others could pound the rod with sledgehammers, driving it as deeply into the rock as it could go. In this case the stone normally split quickly.

A more drastic approach to splitting, when necessary, was to follow the same idea with wooden wedges. The stone was slit with a broad metal-headed tool, into which a series of small, thick, perhaps 4 inch wide, sharpened wooden wedges were forced variously along a desired line, wetted and pounded with sledgehammers. Few Aquia stones were tough enough to reject that intrusion. Whether by stick, rod, or wedge, several efforts might be required. In preparation the quarry master used a metal piece (or his knuckles) to investigate the rock, sounding for its structural integrity before selecting the points of penetration.

The larger, split-off extractions were trimmed into more useful sizes with chisels and smoothed off a bit. Then they were hoisted by a wooden crane and ropes to the stone boat, essentially a wooden barge. From Aquia Creek the barge was towed into Potomac River and poled into the swift and dangerous current of Tiber Creek. Not far up Tiber Creek from the river, a stone landing had been prepared, where the stones from the quarry were removed from the stone boat and lined up on a thick spread of straw over the ground.

Now began a process of selection. Taking great care, the master mason reviewed the "quarry-face" stones in the stone yard before him. Using chalk, he classified the stones with a numbering system, by size and condition as to where they might be modified for use in the walls of the house or for carving. Some stones were moved to an adjacent rough-cutting yard, while others, needing less basic trimming, were towed on a temporary canal that with locks carried them up to about the level of the building site and from there on slides by ox wagon to the construction site itself.[15] A key drawn on paper by the masons identified where various stones might ultimately fit on the walls. They followed Hoban's elevations as far as possible, and improvised. The selected stones were laid out on the ground in the order in which they were projected to appear in the walls. Each was re-numbered. All the time the quarry sap (the water naturally absorbed from the earth) drained in drips.

The stones were given individual identity on paper, one by one, in drawings full scale. Each was laid on the stone it represented and refined as to what shape and size were desired. These newspaper-size drawings—lighter to handle than the stones themselves—were then laid out side-by-side and fitted together puzzle-like, with arranging and rearranging to close up spaces between them. The masons studied the drawings, determining where trimming was called for to make a tighter fit, and then the directions for these alterations, too, were marked on the drawings. Templates were made—like wood frames—for some of the wall stones. After the move toward final accuracy, the individual stones were hoisted on long, heavy wooden tables, where they were chiseled to match the edited drawings. They were then taken back to the adjacent stone yard and laid out accordingly.

Stones reserved for carving were inspected by the carvers, Robert Brown and James McIntosh, recruits from the Masonic Lodge in Edinburgh. They made their own drawings on paper over the stones in the yard, showing the locations of carved decorations, be they garlands or classical borders. They determined which stones had the structural integrity to withstand carved work and placed the carvings accordingly. A single carved decoration might run over several stones, and it must appear cohesive, with no joints readily visible. A simple wooden template was also crafted, but only for sizing stones that would be alike, as the ashlar on the walls—and many of these are not exact. Then these stones were lifted to the tables, where the carving process began. All this careful planning meant that when the dressed and carved stones were finished and placed on the house, everything matched, shoulder to shoulder, top and bottom. It was a well-thought-out, efficient process.

At the Quarry, Splitting the Stone

Splitting the stone was sometimes a three-man job. Here one man holds the "drill"— a threaded wrought-iron spike—while the other two strike the drill with their stone hammers. After alternating strikes, the seated man turns the drill several times, boring as deeply as he can, and the striking resumes. The use of jacks or pointed, threaded rods was more typical, one man applying the hammer. In either case the intrusions in the rock would be renewed in a line until the separation was successful.

The stonemason has inserted nine iron rods or "points" to achieve a split from the larger stone. He taps the points with a small stone mallet while trying to achieve a neat split. Sometimes, depending upon the stone, this split did not take long. Ease of splitting, however, could betray a flawed stone. Aquia quarry is scattered with these.

Upriver, From Quarry to Building Site

An extremely heavy freshly quarried stone is loaded by crane on an oxcart for transport to a waiting stone boat. The stone is yet to be split into smaller building blocks for the President's House. Contact with the stone boat will involve backing the oxcart down an incline and sliding the stone by wooden slides onto the boat. Apparently the crane was used only for loading the oxcart, not for loading the boats.

Loaded with a cargo of heavy stones from the quarry, a wooden boat is poled against the current on the Potomac River to Tiber Creek landing near the building site for the President's House in the City of Washington, some 45 miles upstream. The size of the load is limited by the weight of the necessarily small draft required for floating close to the river's edge, away from the main current's force.

First Stop, the Stone Yard at Tiber Creek

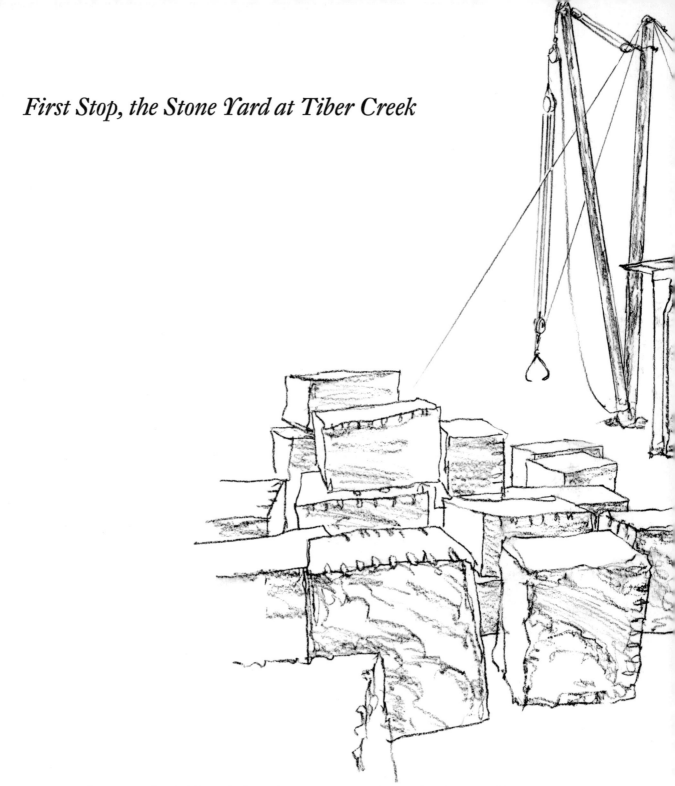

The stone unloaded from the boats bears scars from the process of removal from the quarry outcroppings by splitting with rods and wedges. At this landing the stones are finely graded for quality and, in most cases, further split into smaller sizes conforming for use in the walls of the house. From the Tiber Creek landing the best stone will be moved by canal and ox-slide to the building site.

CHAPTER IX: A BOLD ART

The Finished Carving

IN THE STONE yards on the east side on the construction site, the Scots were very busy with their complex art. Their patterns of work were no different from what they had been in New Town. Nervous efforts by the commissioners led to inquiries about speedy steam-cutting machines, but they came to nothing. Indeed, only this once were building shortcuts considered. It can be said that the entire house was built by hand, just in advance of the new Industrial Age. The bills for tools that the stonemasons asked Hoban to purchase for them kept the blacksmiths of Georgetown busy. Tools were also ordered from Baltimore, most of British manufacture, as was the fine hardware that would be used in the house proper. Rods, axes, picks, smoothing irons, sledgehammers, mallets, and hammers were ordered in the hundreds for the stone lodge. The Scots, masters at their craft, demanded and were given the best.

The most familiar of the stonemason's tools was commonly called the "tool," but formally the "cutter." These were bought or blacksmith-wrought in at least a dozen sizes, essentially a sharpened flat, heavy rectangle attached to a handle. For the President's House the cutter was used primarily for shaving the stone to the desired size and smoothing the outside faces for the walls. The cutter went back and forth over the stone, tilted to suit, until the stone had a sheen, an effect called in its day "polished." Most stones the mason left smooth and polished by vigorous scrubbing with sand and clay before they were taken to the house and set in mortar. Yet another means of smoothing undertaken in the stone yard was with the copper "saw." Worked to divide a single stone into two parts, it thus provided two smoothed faces and two stones by drawing a tight copper wire back and forth, with sand and water continually poured over the incision it made.

Stones that were to be carved were left rough, protruding, and ultimately worked with a number of small tools, as for example the hand-size "tooth tool," a delicate pick-like tool used to create furrowed lines as decorations on the stones. Many sizes of picks were the primary tools, followed by the flat-nosed smoothing tool. Carvers were artists, and individual ones favored different tools, some preferring many and others very few.

The stone carvings at the White House are so richly rendered as to seem to be independent elements stuck on the walls. This is not the case. Each carved feature was cut into the face of the stone to which it belonged and remained a part of that stone. One might say that the carver dug his ornamental element out of the stone, and going deeper, beyond the ornament, terminating at last at the flat background that conformed to the level of the general walling of the house. Some stone carvers, even Michelangelo, imagined that their carvings existed in the stone already, and were merely set free by the sculptor. The Aquia stone was not difficult to carve, and its natural composition was not unlike the denser, less porous Craigleith stone used at New Town. Scottish carving was not delicate but rich and bold, and refined the bulk of the White House. It was a frosting, an embellishment to the block of the house.

George Washington's wish for elaborate carving was fully honored. Just who actually designed President Washington's stone carving on paper is not certain, but the few clues point to James McIntosh. In council with Hoban and his mason colleagues, the details may have been reached in collaboration, but Hoban seems to have turned the carving entirely over to the carvers. The superb carving almost certainly belongs to McIntosh. His details, in which he was likely assisted by Robert Brown, may well go beyond the original intent. Pediment hoods and carved architectural borders decorate all the State Floor windows the same on the north, east, and west sides of the building, rich in carved guilloche or "Grecian chain," lush acanthus leaves, and moldings upon moldings. This was not the shallow carving of New Town but deep carving that had been popular the generation before.

None of the guide drawings or templates used by the stonemasons survive, for typically such materials, thanks to their large number and the wear they underwent in use, were cast out as trash once they were no longer needed. If approached with their preservation in mind, which was unlikely, the workmen might ask, why keep them? Even an architect's drawings might be discarded. Hoban's 1793 revised north elevation for the White House exists today only because it fell at some point into Thomas Jefferson's voluminous papers.[16]

At the Building Site, Finishing the Stone

Smoothing the stone involved many tools, although an experienced mason used only his favorites. Here one man smooths the face of the stone, which will form the outside surface, by rubbing it vigorously with another stone. The second workman uses flat chisels to smooth the surface. The five "backfaces" of each stone would be less fully smoothed.

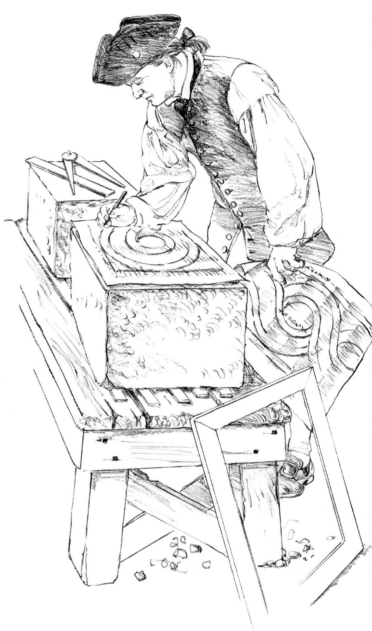

With an iron point of a desired size, the stonemason assures the precision of this carved guilloche border by using a paper drawing. A simple wooden template awaits use in sizing the plain wall stones.

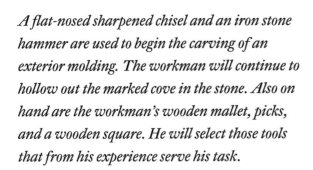

A flat-nosed sharpened chisel and an iron stone hammer are used to begin the carving of an exterior molding. The workman will continue to hollow out the marked cove in the stone. Also on hand are the workman's wooden mallet, picks, and a wooden square. He will select those tools that from his experience serve his task.

Perfecting a solid piece of the heavy stone cornice, the stonemason smooths the fascia with his wooden mallet and iron finish chisel. The block-like tenons at the bottom of the cornice were carved together with the rest and, when turned upright and installed, would fit into mortises prepared to receive them in the walls below. Jack rods and a flat stone hammer await use.

At a "stone table," a mason employs his spike-like iron "point" in refining the volute on one of the heroic pilasters.

A carver with mallet employs a flat-nosed chisel on a developing acanthus leaf window console. Lying nearby is the serrated tooth tool used in every aspect of White House stonework.

CHAPTER X: A SHARED CHALLENGE

Stonemasons, Brick Masons, and Apprentices

THE NUMBER OF masons at work varied from year to year, for in addition to the seven from Edinburgh, other stonemasons had appeared, all, going on names, apparently Scots. Little is known about them, but no complaints called attention to their presence in the records. They wove in and out of the work for short times, apparently paid by the main lodge masons. At some junctures as many as sixteen stonemasons were at work. Other times the payrolls counted twelve, but the payrolls are not dependable for the Scots, who worked on a basis of measurement a little outside the project's mainstream.

Of all the various trades represented on the White House site, the stonemasons worked most closely with the brick masons. And as most stonemasons were familiar with brick masonry, there was a basis of comradeship between the two. Jeremiah Kale, master mason from Philadelphia, brought gangs of brick makers and brick masons to the project. He worked beside the stonemasons building the brick backing of the sandstone walls. Further, Kale and the brick masons helped build the interior bearing partitions and vaults where brick laying was involved. Kale's brick yard north of the building site had several kilns in which he burned his clay forms to whatever degree of hardness the job required, the hotter the harder and most durable. His main production was of soft bricks, tens of thousands of them required for strengthening the structure. Brick that had to face the weather had to be hard, but few of this character were required for backing the stone shell of the White House. The most challenging task the stonemasons and brick masons shared was the structural arching in the basement that would support the very heavy interior developing above.

The structural strength of the outer walls was duplicated inside with arching throughout the basement, giving full support while leaving the floor space largely open for light partitioning into rooms. The engineering techniques of arching and vaulting were ancient ones that stabilized the mass of the White House and kept it firm until the twentieth-century introduction of steel. To envision the original system, think of the house as a large rectangular box, its north-south dimension about 150 feet, substantially longer than the east-west ends, which were about 90 feet. Imagine you look down on this box with the "lid" removed. A regular row of broad and long brick and stone arches springs north-south, lined up parallel, from the front to the back of the house. Further, imagine these "arcades" slashed down the center crosswise by an open corridor, east-west, spaced with heavy piers at each arch. The arcades are connected across this passage with masonry "ribs" of brick that reach diagonally, crowning the squarish spaces in the passage with shallow "domes." The resulting strong arched crossed ribs formed quadrants with ceilings of hard brick supporting heavy loads of rubble, strengthening them, and linking the whole structural system of the lower parts of the house together, drawing pressure from the outer walls and into the interior.

While the north-south arches were ultimately buried in partition walls when the house was finished, the "groin vaulting" of the transverse was exposed, creating a dramatic corridor. The structural beauty of it meant little when the area was service rooms during the first century of the house, but when the basement was formalized as the "Ground Floor" in 1902, the original vaulting lived on painted white. It made a very architectural path through the house. Fifty years thereafter the 1902 corridor was demolished and replicated again in plaster and lath over steel and concrete block.

In theory a perfect stone wall stands by itself with no binding mortar. Building a structure "dry" was the dream of stonemasons, as doubtless it was at the President's House, even knowing the goal would not be realized, for it seldom was—the Mayans notably excepted. In the case of this house the urgency to finish haunted the project, allowing little time to experiment. Thus the stones of the White House were secured with lime-based pulverized seashell mortar, applied thin.

During the course of the actual construction work, many adjustments such as these had to be made by the craftsmen of the building trades. To complete the work with the speed required, they also had to rely on apprentices, although the long-perfected apprenticeship system that both Scots and Irish had known back home proved impossible at the President's House. In Europe, to become apprenticed to an honored tradesman was a privilege, a seven-year learning experience during which the youth lived with the master's family; the "graduated" apprentice would forever honor his master's name. This was rarely the case in America. The successful apprenticeship system that did exist was primarily in navy training, for midshipmen were apprentices, only by another name. These teenagers, usually 14 to 19, were fellows of a higher sort, with a desire to become naval officers. The passion they had was not always shared by boys who apprenticed to learn the building crafts, nor were these boys willing to submit to a strong master. They resented his authority and, even more, the rod with which he duteously bestowed the gift of maturity. And at least on the President's House, the masters found the boys lazy, which was perhaps a more serious complaint even than incompetence. Newspapers in the area abound in advertisements to retrieve runaway apprentices. The usual reward was a cent or half a cent.

The Scots seem to have refused to take white apprentices, and to favor hired slaves. Black men's involvement in the highly skilled aspects of building the White House was not large, but the Scots welcomed them as willing workers, in sharp contrast to most of the other laborers, who had too soft a friend in Hoban. Slaves might find in being hired out for construction an avenue to freedom. Accepting work as laborers or training as building men brought income to themselves and their owners. In the most promising situations, the master split the wages with the slave, allowing the worker to save enough to purchase his independence. A free black might bind himself to a master to eventually earn enough to free a loved one or start a business, of which there were a substantial number in Georgetown and the growing Federal City.

Perhaps half of the African Americans at work on the President's House construction site were enslaved, and free African Americans were involved from beginning to end. Well before the completion of the house, the District of Columbia had among its citizens many free African Americans, individuals and families whose time as slaves, if ever slaves at all, were apparently in distant memory. Records of manumissions in the courts of the region are extensive.

CHAPTER XI: SPRING 1797

President Washington's Last Visit

IN EARLY SPRING 1797, after the inauguration of his successor, John Adams, in Philadelphia, the former president George Washington crossed the country in his coach, with Martha Washington, her granddaughter Nelly Custis, and young George Washington Lafayette, an exile from Revolutionary France. In the Federal City the procession stopped before the President's House. On the outside, it was just about finished. Hoban mustered his militia in a line before the building; they fired a sixteen-gun salute. Washington, apparently not leaving the coach, remained still for a time to admire what he saw and to ponder what it meant. He had never demanded perfection, only completion. The house, he said, could be corrected and adapted as the future saw fit, so he must have been pleased with what he saw. And off he went to ferry down the Potomac, to Mount Vernon. To our knowledge, he never saw the White House again.

The stonemasons finished their carving assignment except in three instances. James Hoban intended to fill the pediment on the north with a splendid carved eagle in a cartouche of sun rays. At some point in construction, however, a decision must have been made not to carve the ornaments, because the stones in that tympanum, or triangular face of the pediment, are perfectly flat, whereas if carved decorations had been intended the raw stone for the proposed carvings would have been left protruding to accommodate it. Conversely, perhaps the omission came when it was decided early to build a portico on the north, which was eventually carried out. It is unlikely that the stonemasons simply left too early to do it. Carving was likewise omitted from stone plaques surmounting the otherwise heavily carved arched windows on the east and on the west. They remain blank.

When the exterior stone of the White House was in place in the fall of 1797, the Scots informed the commissioners that it would not be possible to expose the Aquia stone to the weather. To them this was nothing new. The Craigleith stone of Edinburgh was also somewhat porous. Since time immemorial the Scots had preserved their stone buildings with a thick coat of whitewash. This batter-like compound, made with a base of lime, and added water and flat beer, sealed the walls, filling the cracks and holes natural to the stone. Applied with ordinary straw brooms, the whitewash went on like plaster; it was pressed into declivities. As time passed, rain thinned the surfaces, but the fillings remained, keeping rainwater out, thus protecting stone from being split by the expansion of ice in those crevices. This old Scottish building custom of whitewashing finished stone, if not exclusively Scottish, was put into practice on the President's House in 1798. It did its work. People passing by the huge white structure soon called it, by 1802, "The White House."[17]

What was left to do to the building after 1798 was for the most part the interior finish work, including woodwork and plastering, painting and varnishing, and glazing the windows with their luxuriously large panes of English crown glass. A variety of carpenters, plasterers, glaziers, cabinetmakers, and painters did the work. None lived on the site. The interior can be resurrected only through written clues. The doors were made of mahogany with local holly inlay; on the State Floor they were topped by "cornices." The brass mortise locks, silver plated in the State Rooms, had rings instead of knobs. Window sash was of mahogany varnished. Some amount of cabinetwork was executed by the cabinetmaker Joseph Middleton. In his case we know more because, to the commissioners' outrage, he was using their time for private commissions and lowering his products by rope from upstairs windows nocturnally from his shop in the upstairs oval room. This dishonest act raised a tussle, but the skilled woodworker remained.[18]

Completed for occupation in the spring of 1800, the President's House stood ready for President John Adams on November 1, 1800, ten years after the Residence Act had become law and within the time frame it had established. The grounds were a muddy mess, littered with construction debris and abandoned workmen's cottages. Inside the plaster was still drying, and when Adams ascended to his bedroom that first evening he climbed up the twisting service stairs, as the Grand Staircase had not yet been built.[19] But that the residence was habitable was in itself an enormous achievement. The Capitol, a more ambitious building and of course far larger, was by no means even near completion.

CHAPTER XII: A LABOR OF LOVE

The Great White House

THE PRESIDENT'S HOUSE that today recedes in the dense urban landscape of the capital city was in its beginnings a great naked hulk dominating the barren terrain. To the eyes of the Scots the architecture of the house was familiar, maybe from their apprenticeship days, but compared with the world-famous designs of their own Adam brothers the American palace was maybe fifty years or more out of style. The stonemasons' lives had been touched firsthand by the fashionable designs of the Adam brothers. Collen Williamson, even in rejection, liked to point out his involvement with John Adam, and the Edinburgh stonemasons had built from Adam designs. They learned in their work the Adam brothers' modern classical proportions and taste for limited ornament, so the architecture of the White House, made to please President Washington, was a step backward in proper British taste for the 1790s. Of course the building had started out much larger and more palatial, reduced to house-scale both by circumstances and by Washington's preference.

Washington had greatly simplified the plan from the complex Leinster House model to what one might call an openness typical of American houses. The small corridors, anterooms, and twisting private stairs of the Leinster House plan were omitted in favor of no "secret" spaces. It was a big, accessible house, with what Latrobe called a big belly of an Entrance Hall, from which large crowds could pass along a long Transverse Hall into parlors, a dining room, and a grand public room we know as the East Room. A ceremonial staircase, not built until several years after completion, assured stately entrances. Upstairs was reached by that stair, a steep secondary stair, and a small winding one in a back hall. This chamber floor was also arranged along two sides of a transverse, with smaller rooms.

Carved ornamentation on the exterior of the house is restrained in the sense of outlining certain features, such as windows and doorways. It does not interfere with the smoothness of the stone walls but enhances them, for it is the stone walls one sees from the street and distance. A visitor may approach the White House twenty times and, only by accident, pause to turn and become aware of the high quality of the stone carving in doorways, on the windows, and elsewhere. Seen from the iron fence along Pennsylvania Avenue, the carvings are overwhelmed in the distance by the mass of the house. Only in drawing nearer to the building does one absorb the excellence of the carved stonework of the Scots craftsmen.

By contrast, the base of the house, built by Collen Williamson, is rendered in bold rustication, a configuration of stone blocks around the windows that protrude rhythmically from the flat wall, leaving deep spaces around each block. Especially is this detail notably expanded in the later southward "bow." Here the podium for the South Portico is a heavy pedestal, powerful in the shadows it casts, forming a visual base for the mass that rests upon it. Williamson's work is the distinct flavor of an earlier Georgian vernacular. Beneath the later North Portico, in the deep areaway, is a pedimented doorway that seems to recall former taste, and the detail is unlike any other feature in the White House. The Edinburgh masons' walls rise upon Williamson's basement. Stone modillions encircle the eaves, close-placed, beneath the cornice and the stone balustrade that crowns the house all around. The smooth, flat faces of the north and south walls form plain backgrounds that emphasize the striking stone-hooded windows which seem to hang upon them like pictures, placed symmetrically. Giant arched windows on the east and west fronts stand two full stories up and are handsomely framed in a border of carved fruit and flowers.

The east and west walls provide a flat stone background, as do the walls north and south, yet all but the north side are adorned with heroic pilasters rising full-length, reaching to the roof entablature and cornice, capped by Ionic composite capitals. The climax of these unique capitals is the Double Scottish Rose, carved boldly. In the 1780s this multipetaled rose was propagated in Scotland from native roses, and it instilled patriotic pride there, for it was the rave of gardeners in Europe and the forerunner of the voluptuous Victorian "cabbage rose." The use of single-petal flowers in column caps was traditional in ancient times, but they recede, while the voluptuous, many-petaled roses on the White House announce themselves. The Double Scottish Rose at the White House was unique for its time as an architectural motif and survives there as a signature left by the Scots.

Another episode in the carving program stands out. The Edinburgh masons, in their grandest act, seem to have made a labor of love carving the 14 foot swag above the North Door. Hoban had projected the idea, but the Scots, in designing and executing it adorned this entrance with a halo playfully lush. Double Scottish Roses, acorns, fruits, and oak leaves fill this abundant garland, suspended from fluttering ribbons, all rendered in Aquia stone. It is unquestionably the finest stone carving in eighteenth-century America. One might imagine it a gift to delight President Washington, who inspired the carvers' work.

CHAPTER XIII: A SYMBOL OF SURVIVAL

The Walls Rebuilt

THE SCOTS WERE gone, their services no longer needed by the fall of 1798. Robert Brown decided to stay and opened a marble yard in Georgetown, making tombstones and ornaments. An Alexander Reid, perhaps the same one from Edinburgh, appears in the Capitol construction papers and in subsequent Washington and Georgetown records; the same name appears in the construction records of the Capitol of Pennsylvania at Harrisburg. The rest returned home, took up business again, and re-registered themselves with the operative lodge of Masons in Edinburgh, as was duly noted in the leather-bound tally of membership. Some completed the New Town buildings they had left behind, and there are houses in New Town today documented in deeds as having been built by them. Records of the stonemasons are nearly all otherwise official, for none of their personal writings so far have been found.

For many years after the Scots left, various efforts were made to better the setting of the President's House and create a landscape for it. The muddy declivity behind the house, sloping to the wild currents of Tiber Creek (which still flows briskly beneath Independence Avenue) received some earth filling in Jefferson's administration, creating a terrace on the south side. Jefferson, surely frustrated by the extent of the work and its great cost, planted many seedling trees to cover up the ills of the site. Alongside the federal architect B. Henry Latrobe, he drew up a rough grounds plan, with sweeping drives that defined the public front on the north and the private grounds on the south. His drives and proposed plantings also represented an effort to right the house with its expansive site, which was framed by the city plan to accommodate a larger house. A high retaining wall was begun, and low service wings of stone were built to each side of the house. But nothing was really finished when the British invaded in the summer of 1814 and burned the house down. This act, widely condemned here and in England, was the low moment for America in the War of 1812.

What survived of the Scots' work after the fire? Only the south facade of the house was left entirely intact. The center part of the north front survived, but the flanking walls collapsed, and the two end walls were ruined to the ceiling level of Collen Williamson's basement, which survived totally. Both of Jefferson's wings were lost. On orders from President James Madison, the house was "repaired," and James Hoban came on to do the job. There were few changes, and that satisfied the president, for his objective was political and patriotic, to exactly rebuild the house as a symbol of survival. Madison, like Washington before him, was in a hurry. The next president, James Monroe, would have changed things more, but the project was near completion and he was also in a hurry to move in. He added some fireplaces and partitions, began ordering furniture from France, and listened to Hoban's plans for porches for the house, agreeing to leave the spaces to attach them in the roof with temporary covering.

Few of the original Scots were involved in the rebuilding. Seventeen years had passed since most of them returned home. Traquair was unavailable to help, for he was occupied with his new shop in Philadelphia selling marble busts he made of George Washington, Benjamin Franklin, and the others. One stonemason available for sure was Robert Brown, who was made supervisor of the stonework on the rebuilding. James McIntosh remained at least for a decade, and executed nearly all the carving on the reconstruction. Other stonemasons came and went, most from New York, such as Pat Gill, James Greer, and Alexander Martin, names brought forth because the commissioners literally begged them to stay on the job, yet they all left.

Every effort was made by the superintendent of public buildings, Samuel Lane, and by James Hoban to honor Madison's admonition that the house be merely "repaired." In reports to Congress that word was always used, no matter how many frail walls had to tumble. As much existing stone as possible was reused. Stones with great black licks of soot from the fire, installed in the walls, effectively advertised and defined the work of repair. For new stone, Aquia was again put into service, only now the stones could be carried upriver aboard a steamboat, a welcomed innovation since the house was first built, and raised to the level of the construction site by the Washington City Canal, completed in 1815. When the building was finished the house was painted with opaque lead-based paint over what of the original whitewash remained.

On New Year's Day, 1818, the White House was reopened in a grand reception. Monroe, inaugurated the previous March, was a very popular president. A New England newspaper called his early administration "The Era of Good Feelings,"[20] and the title enraptured the public and the president. Business prospered, and thousands migrated to new land in the West. Many believed that political parties were dead, and turned in admiration toward Monroe, a new executive with a powerful presidential presence like George Washington. Nor was Monroe's stature in any way diminished by his remarkable resemblance to "The Father of His Country." The new president, although republican to his soul, had a diplomat's love of form and elegance and was interested in the house and proud of what it was doing for him. For the New Year's Day opening, he appreciatively ordered a full feast, including wine and beer set up in the basement for the workmen who had built the President's House back again.

CHAPTER XIV: THE IMAGE WE KNOW

Adding the Porticoes

MONROE WAS BEGINNING his second term when the drawings were brought out for porches, planned long before for the north and south fronts of the house. Hoban claimed that Washington had approved porches, but apart from the proposed long terrace on the south, supported by columns, that was never built (although doors from the State Rooms were cut to accommodate it and remain), no one really knew about these other porches Hoban mentioned. Yet drawings did exist, for the Capitol architect, Charles Bulfinch, went to Hoban's house, joined Hoban in taking down from the wall some original White House drawings, framed in glass, and traced the one for a pedimented north porch, which looked about like the North Portico we see today.[21] The openings in the roof beneath canvas covers awaited these new additions.

The president decided to build the south porch. He summoned Hoban, by now a prominent building contractor and entrepreneur in the city, to do the job. Ringed by long, two-story shafts of Ionic composite columns that matched the pilasters along the walls of the north, east, and south sides, the porch was a redefining addition to the house, completed in 1824. It was not in nomenclature an actual portico, for a portico requires a triangular crown, while the porch has a flat roof surmounted by a continuation of the cornice of the house. The new porch, nevertheless, entered history as the "South Portico." The porch on the north was also completed by Hoban, early in the administration of Andrew Jackson. It was in fact a portico with pediment. Had the Scots still been there, its pediment might have boasted the eagle cartouche first planned by Hoban.

Both porches were somewhat odd for the architecture of the house, reducing its scale visually and seeming slightly out of touch with the mid-Georgian style of the house. For this final project in creating the image we know as the White House, Hoban abandoned Aquia Creek for a different quarry upriver, at Seneca, Maryland. In operation since the 1720s, the quarry had sandstone very similar to that from Aquia, but much harder and less porous. Seneca quarry was favored with a water-powered stone-cutting mill, so that cut stones of the desired dimensions could be shipped to the building site. The location was more convenient than Aquia, just out of the Federal District and close not only to the Potomac River but also to the shallow Washington City Canal. Steamboats also shortened the time it took to transport the stone.

James McIntosh, from Edinburgh, was the carver of the columns. He developed the capital design from the pilasters, bringing the Double Scottish Rose into fuller expression on the capitals of the dominant new columns. Over time the stone columns on both porches have suffered relatively little wear, which may assure that they are of tough Seneca stone. Yet a quarry document suggests that they may be Aquia.[22] The portico on the north bears an especially heavy burden in the full stone execution of the entire pediment its six columns support. Seneca stone at that time was undergoing a change, as the quarry workers dug deeper into it. The near match to Aquia stone was suddenly shifting to the dark bloody color cherished as "Washington brownstone" and later used to build the Smithsonian Castle on the Mall.

CHAPTER XV: REBORN OF STEEL

President Truman Saves the Walls

THE HOUSE STOOD as it was for a century and a half. Monroe's modern interior had been rich and European, with gold-leaf–covered chairs and stools, a great Napoleonic chandelier, and silk hangings. Subsequent renovations for many years largely shifted furniture and replaced worn-out carpets. Yet modern improvements—plumbing pipes, gas pipes, electrical wiring, chases for heating pipes, radiators, ventilators—often gave unseen torture to the stonework. Radical but superficial physical renovations in 1902 by Theodore Roosevelt not only did not solve the problems but increased them with quick-fix construction. In 1927, the attic was turned into a third story, with a floor of concrete and steel tucked away beneath a new, low-pitch roof that concealed the addition from street view. The tremendous weight of this new feature literally mashed down upon the Scots' walls.

The week after the attack on Pearl Harbor, December 1941, brought the Army Corps of Engineers into the house to inspect its structural condition. So much was made of wood, and inside the stone walls the brick and wooden structure had sagged. Plaster hung on wood lath. The wood floors of pine, oak, and mahogany, with their large pine supporting timbers, were still tightly packed with bug-proof cedar shavings between the floors and the ceilings below them. Danger from fire was everywhere. It was estimated that one firebomb, appropriately placed, might turn the house into a ball of flames. President Franklin D. Roosevelt was not troubled by the resulting report. He judged it as excessive and, loving old houses himself, ignored most of it. Nor was he moved by the suggestion of Civil Defense that the house be painted camouflage on the outside.

Yet the report was still on the table when President Harry S. Truman took office, and the Army Corps of Engineers made sure the new president saw it. Curious occurrences in the house that might otherwise have passed unnoticed began to spice the narrative. Chandeliers in the Blue Room and East Room began to sway slightly. Plaster dust descended snow-like from cracks in walls and ceilings. When Margaret Truman, the president's musical daughter, loaded her Second Floor sitting room with several heavy pianos, the leg of one of them slipped between two floor boards, sending chunks of plaster ceiling falling into the room below. This event made headlines, and it was evident that something had to be done. In 1948 President Truman and his family moved out of the White House and into Blair House, the guest house Franklin D. Roosevelt had created in 1942 from an old family home across Pennsylvania Avenue. For the next three and one-half years the White House was under renovation, the only radical change that had come to it since the British burned it in 1814.

So adamant were the engineers and Secret Service against the house for its structural weakness and threat to the president's safety that the first solution was to tear it down to the ground and rebuild a new house that looked like the old one. Unquestionably this would have taken place had President Truman's sensitivity to history and its symbols not been so powerful. He simply stopped the idea in its tracks. What was the least that would have to be taken away to make the White House serve the modern purposes of an official house? Fireproof construction, up-to-date amenities—all these the president approved—and he knew they could be handled without a total demolition.

Truman learned that at Yale University the venerable eighteenth-century academic building on campus, Connecticut Hall, a 1750 brick structure of great historical interest, had been found unsound and was under restoration by the architect Douglas Orr. The plan being followed was to strip the entire interior, save the old brick walls, and rebuild the interior in modern materials. This work was conceived a few years before and in process as the future of the White House was being considered. In conference with Orr and Lorenzo Winslow, the White House architect, President Truman adopted the Yale plan. A committee of architects worked out the details and a committee of Congress approved. This plan was, of course, the salvation of the stone walls the Scots had built in the 1790s and why they still stand today.

In 1948 the dismantling of the interior began. The walls of wood, plaster, and brick, the board flooring, and the trim were attacked with pick and sledgehammer. Winslow, an antiquarian, put aside what he could to reuse. Elevator-like steel-framed shafts were introduced through the house, from 12 feet below the basement all the way up to the base of the 1927 Third Floor, which they supported like table legs. Through these shafts the debris was poured out windows into dump trucks, destined to become landfill in Fort Myer across the Potomac. At last only the shell of stone remained; the brick backing of the stone was entirely removed. And the old walls stood solidly on their own.

The president was intensely interested. When in town, which was most of the time, he made a daily tour of the works. One day he saw men with air hammers preparing to widen a basement door to give access to a bulldozer and a dump truck, for the purpose of digging the new subbasements. These cellars were a secret add-on to the budget, intended for domestic storage and machinery but mainly to be called for when needed for security. Truman knew well why the air hammer men were there, but he stopped them. There was to be no widening of anything because it cut into original walls. To obey the presidential edict, both vehicles were dismantled and rebuilt inside the shell of the house. When the desired level was dug out and the soil hauled off, the ground was far enough under Collen Williamson's foundations for both machines to escape intact. Thus the subbasement was excavated without damaging the underpinned stone. Survival of the Scots' stone walls was President Truman's personal prerogative. Left to anyone else involved, they would have been destroyed. Truman's rescue was not antiquarianism. He saw the walls as American symbols distinguished by their creation on order of George

Washington. They were touchstones to the founding of the nation and, if preserved, would maintain the historical validity of the house. It was less a matter of what they meant than what they were.

When digging and removal ended, there was a brief time when the insides of the stone walls could be examined. The skeleton of steel that would form the bones of the house was in place, including the shafts, which would remain, for they were part of the reconstruction plan. It became interesting to the president and workmen that the backs of a great number of the stones, facing inside, had marks cut into them. Truman declared them "Masonic" symbols. This was maybe half true. They were in fact "banker-marks," the trademark signatures of the Scots stonemasons who cut them into the stone, the concluding strokes in the process of measurement. They were signals that it was time for payment. The sudden reappearance of these marks from beneath a thick layer of brick verified the Scots' extensive use of measurement for payment in the stone construction of the White House. The president ordered a sufficient quantity of individual marked stones taken out of the wall and sent one with his greetings to each chapter of the Masonic Order in the United States. There they reside today, treasured, with marks in some cases identifiable to the Scots who built the White House.

The interior was reborn in steel, "of skyscraper strength," and concrete tile, following, in most of the building, the same plan Hoban had made for George Washington. Wartime fears of danger were wiped away by a revised house considered fireproof and bombproof. Modern conveniences were tucked in. Removal of several feet of soft brick that had been built up against the back of the stone walls gained space for such necessary improvements as air ducts, closets, and the like, without intruding on the original dimensions of the principal rooms. The steel and concrete structure within did not touch the Scots' stone walls. New footings strengthened them, and steel spines, where necessary shored them up.

When President Truman reoccupied the house in late March 1952, the White House looked as it always had on the outside. Within it was a place that felt different; one sensed its new steel strength when entering the building. Little repair was made to the stone walls that still composed the White House from the outside. Pocks and cracks were covered over with fresh white paint in several coats. The house looked the same because the walls were the same. Beneath their sheen of white, they were irregular in places, giving a ripply effect, like that of sheets blown out stiff on a clothesline. A modern surface would have been as slick as glass. Nothing about the walls fell into that category. There were windowsills chipped at the corner, carved roses with missing petals, and many other scars left by time.

CHAPTER XVI: THE HOUSE MUST BE PERFECT

Preservation of the Stone

TWENTY-FIVE YEARS AFTER Truman's reconstruction, Rex W. Scouten, chief usher of the White House and thus the first executive officer of the compound, observed that the exterior walls would no longer hold paint for more than a year. Keeping the outside looking good was a continuing challenge, not to mention cost. He had special interest in the historic house and had been present as a Secret Service officer when the Truman renovations took place. To oversee the White House, as he did for many years, was to address every detail involved, from official ceremonies to daily life. As the home of the president, the house itself must be perfect. He had inherited President Truman's determination that the walls be preserved.

Scouten brought the problem to President Jimmy Carter, also a student of history, with a plan he and officials from the National Park Service, legal stewards of the White House and grounds, had developed by talking to stone experts. The conclusion was to strip the walls rather than to continue painting. President Carter gave approval to begin the work. Scouten's assistant usher, Gary J. Walters, and James I. McDaniel from the Park Service were put in charge. It was a big, messy job that ran on for twenty years, through four presidents. Some forty coats of paint lay on the walls. The first, put on by President James Monroe, covered the Scots' whitewash, and it was the whitewash that proved the hardest coat to clean off. It had performed to perfection its job as a protective covering and filler of holes in the stone. A special chemical paint remover was invented to penetrate the white layers of nearly two centuries.

The White House was stripped. The stones stood bare, and the carvings appeared in sharp relief. Then followed a long process of repair, involving replacing deteriorated stones and making special new patches of Aquia material. A few decorative elements were recarved. Rusty wrought-iron supports, L and T brackets made originally to bind some stones to the brick behind them, were either removed or replaced. The earliest untouched section of original wall was found above the South Portico ceiling, covered up since 1824.

The White House was cleaned and repainted, using brushes but also air-spray machines. The old stones shimmered beneath the new white paint (Duron Paint Company's "Whisper," Number 2486W) softened by a slight yellow cast that seemed entirely white in contrast to its setting of greenery. President Bill Clinton accepted the final work in 1996.

Notes

1. Principal sources for this essay are documents of the building of the Federal City, located in Record Group 42, United States National Archives Building, Washington, D.C.; records in the Masonic Lodge, Edinburgh, Scotland; records of the City of Edinburgh; and records in the National Archives of Scotland. In addition I made extensive use and interpretation of the remains of Aquia quarry and the walls of the White House itself. Other support sources are found elsewhere in these notes.

2. An Act for Establishing the Temporary and Permanent Seat of Government of the United States, July 16, 1790, online at the Library of Congress website, A Century of Lawmaking for the New Nation: U.S. Congressional Documents and Debates, 1774–1875, www.memory.loc.gov.

3. Thomas Jefferson was clearly dubious about Washington's ability or the value of his opinion in the planning and building of monumental buildings. The secretary of state was very guarded in his criticism, and political besides. Washington, too, seems to have been dubious about Jefferson's own architecture, from the odd, unfinished Monticello, where pictures hung on raw brick walls, to the very odd, if innovative Virginia State Capitol then beginning to rise in Richmond, patterned on a Roman temple. Both were architecture buffs, both liked to plan and build. It is difficult to imagine the Georgian-based Washington and the modern neoclassicist Jefferson as restrained rivals, when laid before them was an opportunity of a lifetime to influence ambitious building.

4. For more on Hoban, see the special issue of *White House History* devoted to him, no. 22 (Spring 2008), and Andrew McCarthy, "Architectural Investigation and a Neoclassicist Rediscovered: James Hoban's 1792 Designs for the President's House," *White House History*, no. 42 (Summer 2016): 16–33.

5. See William Seale, *The President's House: A History*, 2nd ed. (Washington, D.C.: White House Historical Association, 2008), 1:57–58. The original contracts are found in Record Group 42, National Archives, the earliest beginning with the building of the White House.

6. The history of Williamson, Adam, and Sir Ludovick at Moy House is found in the extensive Grant Papers, in the Seafield Estate Papers, on deposit in the National Archives of Scotland, Edinburgh.

7. Commissioners of the Federal Buildings, to Collen Williamson, August 29, 1792, Commissioners Records, Redord Group 42, National Archives. See also the account of the cornerstone event in Seale, *President's House*, 1:36–37. In spite of many twentieth-century searches in the walls, no trace of the cornerstone has survived.

8. Payment accounts and payrolls appear all through the Commissioners Proceedings and Records, Record Group 42, National Archives. See also Robert James Kapsch, "The Labor History of the Construction and Reconstruction of the White House, 1793–1817" (PhD diss., University of Maryland–College Park, 1993).

9. Williamson to Commissioners, June 5, 1794, March 11, 1796, and to President Jefferson, June 11, 1801, Commissioners Proceedings, Record Group 42, National Archives.

10. Williamson to Jefferson, June 11, 1801, Commissioners Proceedings, Record Group 42, National Archives.

11. Collen Williamson, estate inventory and codicil to will, July 1, 1799–August 22, 1807, District of Columbia Court Records, Record Group 42, National Archives.

12. Lady Margaret Grant to Lady Pitcalnie, Edinburgh, 1788, Seafield Estate Papers, National Archives of Scotland.

13. James Traquair to Jefferson, Philadelphia, [November 30, 1792], *The Papers of Thomas Jefferson*, vol. 24, 1 June–31 December 1792, ed. John Catanzariti (Princeton: Princeton University Press, 1990), 683, online at National Archives, Founders Online, http://founders.archives.gov.

14. Record and tally book, Masonic Lodge Archives, Grand Lodge, Edinburgh.

15. The commissioners signed an agreement with Patrick Whalen on September 1, 1792, to build a canal 15 feet wide, a diversion of water from James's Creek to Tiber Creek, to have earth walls, to be 2 feet deep at low tide, free of stumps. Commissioners Records, Record Group 42, National Archives.

16. The two significant drawings by Hoban of the White House at the time of its construction are a plan and section of what he seems to have originally proposed and an elevation of the house from the north. The first, in Jefferson's papers in the Massachusetts Historical Society, shows the original three-story proposal, while the elevation, in the Maryland Historical Society, shows the final compromise from the fall of 1793. This plan of a year later than the first gave various details to the stonemasons, but there would have been others, notably three additional elevations and almost certainly some details. None of these survive. See Susan E. Smead, "Hoban's Design for the President's House" (MA thesis, University of Virginia, 1989).

17. Donald R. Hickey, "When Did the White House Become 'the White House'?" *White House History*, no. 41 (Spring 2016): 4–11.

18. Seale, *President's House*, 1: 74–75, 76–77. Full documentation of the cabinetmaker embroil are in Commissioners Proceedings, 1798–99, Record Group 42, National Archives.

19. William Seale, "The White House in John Adams's Presidency," *White House History*, no. 7 (Spring 2000): 26, 29.

20. *Boston Evening Sentinel*, July 12, 1817.

21. The drawing by Hoban hung framed under glass on a wall in his house. Presumably there were others, but the porticoes had been of great interest, and this one seems to have had special meaning to the architect. It was lost in a fire shortly after Hoban's death in 1831. The only record of it is the tracing made by Bulfinch, preserved in the Records of the Rebuilding and Repair of the President's House, Record Group 42, National Archives.

22. A drawing by Thomas Towson that shows how Aquia stone columns might be cut from a single outcropping is preserved in the Commissioners Records, National Archives, along with a letter dated July 1, 1824, from Towson, manager of one of the quarries of Aquia stone.

Catalogue

CHAPTER I

Aquia

IN 1608 THE EUROPEAN explorer Captain John Smith mapped the Potomac River and its tributaries, negotiating trade with the Patawomeck Indians who were prevalent in the area. Smith traveled up Aquia Creek, twice referring to it in his papers as Quiyough (kwee'yuh) (Algonquin for "gulls")—once when describing the creek and again when naming the Indian village on its south side.

In 1647 Giles Brent established the first English settlement along Aquia Creek on a 17-acre island, originally named Brent's Island or Wiggington's Island. Nearly fifty years later, in 1694 George Brent (Giles Brent's nephew) became the island's first documented owner. The property remained in the Brent family for almost one hundred years, during which time it was used as a private quarry. The island quarry was a rich source for what we now refer to as Aquia stone. It was a "freestone," so named because it could be freely carved without splitting. This property combined with its beautiful white color made Aquia stone a highly desirable building material.

In 1791, the federal government purchased Wiggington's Island to provide stone to build the President's House and the United States Capitol in the new City of Washington under construction on the Potomac River. From 1791 through the 1820s extensive quantities of freestone were extracted from this site. In 2010, the island was acquired by Stafford County, Virginia. Now called Government Island, the site of the former quarry is now a public park with trails and markers highlighting its historical significance.

This detail of a 1670 map of the middle reaches of the Potomac River shows the locations that figure in this book from the Aquia quarry (marked "Brents" and circled on the left) eastward to the site of the new Federal City (marked "Turkey Buzzard Point" and circled on the right).

Drawn in 1818 during a visit to the quarries around Aquia Creek, this ink and wash sketch by B. Henry Latrobe captured a rock outcropping after the intense harvesting of it was over. Latrobe (opposite, top) included labels and measurements on the drawing.

Although in use at various junctures after 1817 and largely abandoned since the 1820s, the Aquia quarry on Government Island as it remains today (above) almost repeats Latrobe's sketch in its appearance.

These recent photographs made at Aquia quarry amid the autumn leaves of woods that have grown around and upon the outcroppings of sandstone include surviving examples of eighteenth-century splitting. The primary means of removing the first slabs from the natural outcropping of rock, splitting worked well in the initial reducing of natural stone to a smaller size for use in building. The advantage was that the result provided two relatively smooth faces. The resulting stones were always too heavy for a man to lift. In the split sections shown here, the desirable stone has been harvested, leaving the natural outcropping from which it came. Clearly, once they could inspect it inside, the quarrymen were not pleased with the quality of the pictured outcropping, for they left it behind. The slice they removed, however, could have been cut into smaller blocks, or itself cast aside as imperfect.

Discarded stones are scattered throughout Government Island today, having gone through the harvesting and shaping process but at some point having failed the test. The stonemasons, and especially the quarrymen, constantly evaluated the material during the transformation from natural rock to building block. Intuition had much to do with it. They would have tapped the stones with iron and wood, or even pounded knuckles on them, "sounding" for imperfections. The stones pictured above and opposite have seen considerable work toward becoming building stones, but their journey ended for some physical defect.

This tumbled spread of stones was given shaping time by stonemasons, only to be discarded. Note the chisel marks of early splitting and smoothing.

In 1791 the United States government purchased all but one acre of Wiggington's Island. That one-acre parcel had been sold previously in 1786 to Robert Steuart, a stonemason from Baltimore, Maryland. Steuart delineated his parcel with four boundary markers made of stones much invaded by tests with chisel and rod, and left unused. The largest of the markers still remains on the island and is clearly marked with his initials, "R · S."

After the quarrymen split off the best of this outcropping, they left the rest. Nature followed over time with its invasive tree roots, verifying the quarrymen's judgment that the section of stone at right, with its holes and recesses, was too weak for the building project. Note the cuts and attempted cuts of the mason's long-handle smooth-nose chisels, as well as the rod-like jack driving tools. Tracks of tools of many kinds can still be read in this rejected rock mass. A part of it has been sheared off, part of it transformed perhaps into building blocks. Nature has a strong and destructive presence in tree roots, which can split a rock as powerfully as hammered jack tools, rods, or wooden wedges.

The lichen-covered outcropping of natural stone above was split successfully into two parts, yet left unused. The stonemasons would trim, or "quarry-face," the six faces of the stone with their chisels (as in the example seen opposite) before further cutting the stone into smaller blocks. Quarry-face is the first step to ridding the forward side of the stones of excess. It was a popular final face for stones in the 1870s through the early twentieth century, but it would not have pleased President George Washington for the White House. The next step was to smooth, or make the "finish-face," of the side to be exposed in the walls of the building to come. The remaining five faces, which were to be buried in the walls and unseen, were left unsmoothed.

Aquia stone was very heavy—120 pounds per cubic foot—and had to be transported to the site of the White House by way of wooden boats that were loaded with stone dragged by oxen to the wharf, now long gone from the area pictured here. An inclined ramp at the creek bank accommodated their slide from wagon to boat. Aquia Creek was dug out into a kind of bay and given a cut some 18 feet deep. All the quarries along Aquia Creek were served by this large water area. In the swift current, the boat followed the curving path of the creek, entered the Potomac River, and continued on to the Federal City. This now tranquil area is surrounded today by leafy residential suburbs but was a busy, crowded place through the 1790s, with several hundred men at work at times. There were rows of small houses to accommodate them, a blacksmith shop, tool houses, stables, and a great kitchen where cooks labored night and day to feed the workers. After 1800, most of it vanished, a deserted village that rotted away with time.

One of the best examples of the early use of Aquia sandstone is the mid-eighteenth-century wall of the Masonic Cemetery in Fredericksburg, Virginia. The wall was very possibly built without mortar, for the mortar one sees today is applied, and certainly a much later safety application. To lay a wall without mortar is an ideal in stonework and a stonemason's dream. This wall, for all its beauty and excellence, shows the vulnerability of Aquia stone to the weather.

Built as one of a line of "public chapels" by the Church of England, Christ Church has remained in continuous use since its completion in 1773. Its fine brick walls are framed by restrained decorations carved of Aquia sandstone. The designs of the tripartite window over the altar (opposite) and the enriched door surround (above) also appear later at the White House. Both examples are found in James Gibbs's 1829 Book of Architecture. *Paint has protected the Aquia quoins and window keystones from the weather for more than two centuries, yet the brick, less susceptible, remains bare.*

Built by the patriot George Mason in the 1750s in Fairfax County, Virginia, Gunston Hall, resembles smaller country houses of Ireland. It had already stood nearly half a century when the White House was built. While the Aquia stone quoins at Christ Church in nearby Alexandria, Virginia, have been protected, like the White House, with paint and show little damage, those at Gunston Hall, near duplicates, have been subjected to the weather and show their age.

CHAPTER II

Enter James Hoban

THE NARRATIVE OF AQUIA stone's application to the White House describes the fulfillment of George Washington's wish that the President's House be a house of stone. James Hoban's design (opposite, top) and his specifications (opposite, bottom) were, after the decision to use stone, the next major steps. The design for the North Front was accepted by Washington and the Commissioners, but it was not Hoban's first. A modification of his original concept, the design is one story shorter and enlarged by one fifth. Dated October 30, 1792, Hoban's preliminary set of specifications marks the beginning of the building of the White House. The letter reads, "In consequence of your appointing me to superintend the Building of the President's House, I have made a calculation of the principal materials that will be wanted for that Building, which I shall submit to your consideration, to point out the mode, that to you, may appear proper to collect them. They are many of them expensive articles and will require much care and attention to the choice of them, being much out of the common line of Building as you will perceive from the inclosed list." The following pages reproduce the original document, now in the National Archives, in its entirety.

The Commissioners of the Federal District

Gentlemen

In consequence of your appointing me to superintend the Building of the Presidents House, I have made a calculation of the principal materials that will be wanted for that Building, which I shall submit to your consideration, to point out the mode, that to you, may appear proper to collect them. They are many of them expensive articles and will require much care and atention to the choice of them, being much out of the common line of Building as you will perceive by the inclosed list —

Washington
Oct.r 30.th 1792

I am with great respect Gentlemen
Your most Obedient Serv.t
James Hoban

2 Compleat setts of Blocks Iron Strapped and Metal Shieves with falls compleat to hoist 2,000 weight 53 ft high

2 Do. to hoist 1,000 weight — — — — — 53 ft high

5,00 fathom of 2½ Inch Rope tared

5,00 fathom of 4 Inch Rope tared

10,000 feet of two Inch plank for Scaffolding, gangways &c.

2,00 Scaffold poles

1,00 Putlogs } for Scaffolding

3,00 Legers

300 Logs of White Oak or Yellow Poplar for Girders, Joist Roofing &c
  50 of them to be 42 feet long from 18 to 20 Inches Squr
  50 — — — — 40 feet Do — — — 16 to 18 In Squr
  50 — — — — 35 feet Do — — — 16 to 18 In Squr
  50 — — — — 30 feet Do — — — 16 to 18 In Squr
  100 logs of such lengths as can be got from 20 to 30 feet

80,000 feet of them clear white Norward Pine from Casco Bay or any part Eastward of Boston — one half of Inch, ¼ of 1½ Inch, and ¼ of two Inches thick

60,000 feet of Flooring Plank from 5 to 6 Inches wide 1½ Inches thick free from Nots & Sap

N.B. Mr. Bowmans terms is 14 Shillings Sterling pr. Hundred, Mr. Bennetts is 15 Shillings, and the Longest lengths that can be got of the Quality wanted, is from 24 to 30 feet. The sawing of Plank here by hand, will come at about 11/6 this money pr. Hundred. This article requires to be particularly attended to. Mahogany on the way

80,000 Slates, but may vary according to the goodness of the Slate so as to make the Number less. a merchantable Slate shewing to the weather 24 Superficial Inches

Hoban's specifications include the supplies needed for flooring, scaffolding, joists, and girders.

10 Tons of Iron for Cramps Straps Bolts nuts Screws &c.
20 Tons of Plaster of Paris for inside finishing
10 Hundred weight of Spanish Whiting

N.B. Paints of different kinds will be wanted, as, White Lead, Spanish Brown, Yellow, Oker Lamblack &c and Lintseet Oyle, but being articles liable to much waste by soakage &c, I would recommend to get the materials in the gross, and to put up a Mill to grind the colours that will be wanting on the Spot, which will be a saving of 30 or 40 p Cent —

Nails

150,000	of 4	Nails for Lathing
50,000	6	Nails
150,000	8	Nails including 90,000 Clouts for Slating
150,000	10	Nails to be used for flooring &c
50,000	12	Nails
150,000	24	Nails for Roofing &c
15,000	30	Nails
15,000	40	Nails

Brads

50,000	small 2	Brads
150,000	3	Brads
150,000	4	Brads
150,000	6	Brads

20 lb of Needle Points

20 Quire of Patent Glass paper 1st 2nd and 3rd Quality

Hoban also requests many tons of iron and Plaster of Paris and thousands of nails and brads in various sizes.

On November 3, 1792, James Hoban again wrote to the Commissioners of the Federal City concerning materials needed and also suggested that he invite stonecutters he had known in Ireland to come across the sea to work on the new buildings. He begins, "Being universally acquainted with men in the Building line in Ireland, particularly with many able Stone Cutters in Dublin with whom I have been concerned in building . . . to those men I could write

Oak, Yellow poplar, or Cypress, the last being equal to any for its durability, I would wish that something may be done at this meeting respecting this article, as the Season for geting it is approaching fast.

Mr Cabots information respecting the plank at 13 dollars ⅌ M is Reasonable, but it must be observed to him, they must be clear plank; what is understood by clear plank is, not three nots in a plank as large as a man can cover with his thumb, which is the Quality that I have described and expect to get, common he rates at 10 Doll^{s} ⅌ M

With respect to the slate from Boston at 6 Doll^{s} ⅌ Sq, it comes nearly to the price of Scotch Slate at Chatton, at five Guineas ⅌ M, one thousand of which has covered three squares, about £1..16..3 ⅌ Sq^{r} and the scotch Slate has the advantage, by more than the difference in price, in the goodness of its Quality.

Nails being constantly in demand here, I request the Commissioners will adopt some mode of geting a temporary supply, as what we have got is at an advanced price, and no regular method of being supplied.

I would also wish, to have orders left, that if any Seasoned Stuff should come to Geo. Town, or coud be procured elsewhere to have it purchased, that we may be enabled to proceed with such part of the work as it may suit, and am with great Respect Gentlemen Your

Novr 3^{d} 1792   Obedient Servt Jam^{s} Hoban

if it meets the approbation of the Commissioners, to embark for the City, early in the Spring, and hold out such terms to them as the Commissioners may think proper." Hoban continues his letter with details about the need to obtain stone, lumber, and nails. As far as known, no Dublin stonemasons came to join him, however many an Irishman worked building the White House. Hoban was always their defender and promoter.

CHAPTER III

It Is Agreed

COLLEN WILLIAMSON'S AGREEMENT with the Commissioners made August 29, 1792, was a rather general arrangement that he find, hire, and pay the stonecutters and masons. He was to be entirely over that department of the work, while being available to participate if necessary in other aspects as well. His contract, now housed in the National Archives, reads: "It is agreed between Colin Williamson and the Commissioners of the Federal Buildings that the said Colin Williamson will superintend the Stone Cutting in the City of Washington for the public Buildings . . . and the laying the same stone and such Part of the Masonry as his Attention & Skill may be thought necessary or useful in and will afford his Assistance in the hiring Stone-Cutters and adjusting their Accounts and in general in promoting and conducting the work—For which Services the said Williamson is to be paid four Hundred Pounds Maryland Money . . . and if any travelling expences should be incurred by Journies to the Stone Quarries." Williamson was the first stonemason taken under contract and he and the lodge he assembled began construction of the White House.

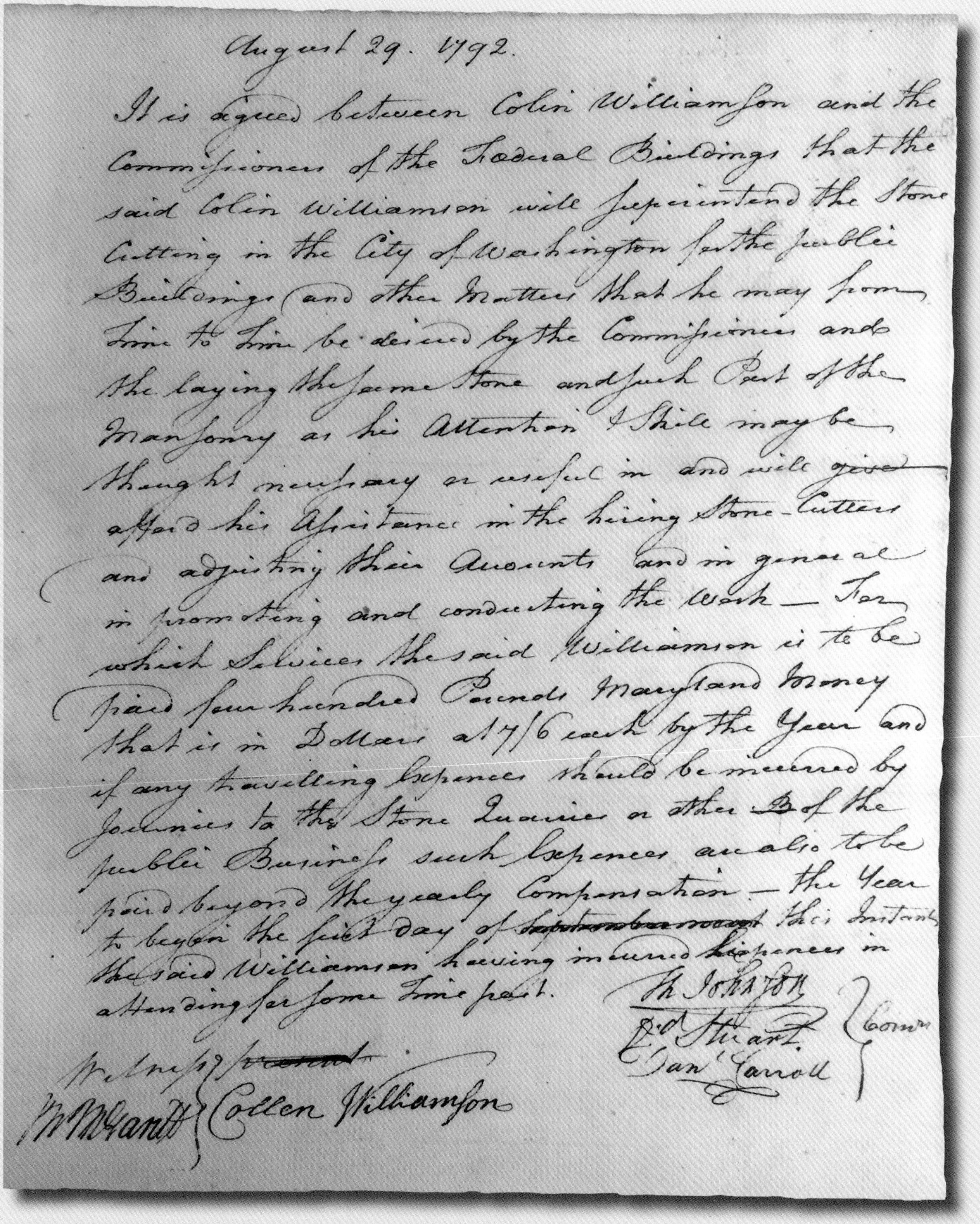

August 29. 1792.

It is agreed between Colin Williamson and the Commissioners of the Fœderal Buildings that the said Colin Williamson will superintend the Stone Cutting in the City of Washington for the publick Buildings (and other Matters that he may from time to time be desired by the Commissioners) and the laying the same Stone and such Part of the Masonry as his Attention & Skill may be thought necessary or useful in and will give afford his Assistance in the hiring Stone-Cutters and adjusting their Accounts and in general in promoting and conducting the Work — For which Services the said Williamson is to be paid four hundred Pounds Maryland Money that is in Dollars at 7/6 each by the Year and if any travelling Expences should be incurred by Journies to the Stone Quarries or other B of the publick Business such Expences are also to be paid beyond the yearly Compensation — the Year to begin the first day of ~~September next~~ this Instant the said Williamson having incurred ~~Expences~~ in attending for some time past.

Witness present
Wm McCants   Collen Williamson

Th Johnson
Dd Stuart   } Comrs
Dan Carroll

In eighteenth-century Scotland and England, working stonemasons granted bankers-marks to apprentices upon completion of their training. These marks represented the identities of the newly trained masons, and in many cases were variations of the teacher's design. The symbols provided a history and background to any future employer and were registered and protected by the mason's guilds or lodges. The marks served a practical purpose, indicating that the stonemasons' work was paid for by "measurement," not wages. The practice was introduced to the White House construction by Collen Williamson to whom it would have been as natural as breathing. His Commissioners, committed to wage pay by the day, disliked measurement as inefficient. Along with the industrial age, they won out long after the White House was occupied. The old custom, now abandoned, reaches back into the earliest civilizations. Most of the White House bankers-marks are carved on the back or on hidden surfaces of the building stones, unseen until stones were removed during President Truman's renovation of the Executive Mansion in 1948–52. Some of these stones were distributed by Truman to state and other Masonic Lodges in North America, while a number of them were retained and the marks displayed in two reconstructed fireplaces on the Ground Floor of the White House.

85

CHAPTER IV

The Finest Stonework in America

THE AQUIA SANDSTONE WALLS of the President's House, built to James Hoban's design and completed by the skilled Scots, rose as George Washington intended. They boasted the finest stone carving in America at that time. The walls survived the wartime fire of 1814 that gutted the interior and stood throughout the Truman-era renovations. By the 1970s, more than thirty coats of paint preserved the stone but also obscured the fine carving. The removal of the paint and preservation of the walls that began in the Carter administration and continued for more than twenty years allowed for thorough study and documentation of the stone masons' work. This stone-by-stone to scale drawing of the North Front of the White House undertaken by the Historic American Buildings Survey of the National Park Service from 1988 to 1992 was made to include the grain pattern of the Aquia sandstone. The scholarly investigation welcomed inquiry—HABS architects, for example, produced this north elevation as it would have appeared before 1830, before addition of the Hoban/Bullfinch portico. Drawn without the North Portico, this is the house as it would have been known to John Adams, Thomas Jefferson, James Madison, James Monroe, John Quincy Adams, and for a time Andrew Jackson. The walls, most of "ashlar" (uniform blocks), occasionally defy this rule, as HABS has duly recorded. From this vantage point we see the monumental stone swag above the North Door, the pilasters, window surrounds and hoods, balustrades, modillion cornices, and ionic capitals.

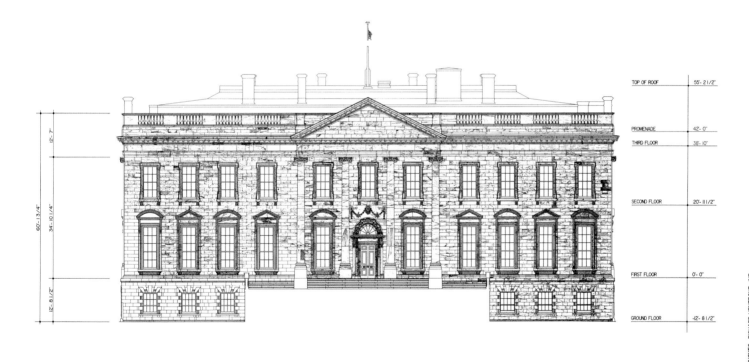

NORTH ELEVATION — SHOWING GRAIN PATTERN OF AQUIA SANDSTONE

THROUGH AREAWAY; FRONT COLONNADE REMOVED
NOTE: STONES WITHOUT GRAIN PATTERN WERE PAINTED OR INACCESSIBLE AT TIME OF RECORDING

FEET 1/8"= 1'-0"
METERS 1:96

MATERIAL NOTES
1. WALLS - PAINTED AQUIA SANDSTONE FACING ON BRICK
2. SASH AND DOOR - TYPICALLY WOOD; GROUND FLOOR WEST WINDOWS, METAL

Splendid carving around the North Door and the surmounting swag seem to honor George Washington's insistence upon carved ornament for the President's House. The climax of the fine stone masonry is this garland that crowns the main entrance. In about 1796, the draped festoon was hand-cut into the faces of two great Aquia stones, which, when finished, were hoisted with pulleys and rope to their present location and mortared together side-by-side, 14 feet across. Fruit, roses, acorns, and oak leaves, gathered by ribbons, are worked into both garland and door surround below with acanthus leaves, pearls, and medallions that match the interior door trim visible inside the house. Central to the scheme are the Double Scottish Roses, which play merrily, in full physical detail from the oak leaf swag.

The fashionable Double Scottish Rose, first propagated in Scotland in the 1780s, was a great source of pride to the people of Scotland. Adopted by the Scots stonemasons as a motif for the decoration of the President's House, it was introduced in about 1796 on the heroic pilasters of the east and west ends and on the south front of the house. It also appears in the garland on the North Front (detail opposite), and was repeated in the portico columns of 1824 and 1829–30. The above engraving, found in the collection of the Oak Spring Garden Library, was made in 1798 at a time when it was one of the most popular roses hybridized in Europe.

The heroic pilasters that ring the house on three sides meet with logical success on the corners of the building, giving the continual flat walls their own corner, while not losing the power of the pilasters. To study this corner solution on the southeast (which has matching treatment on the southwest) and its cornice is to appreciate the skill of the builders.

The pilaster caps boast acanthus leaves caressing the Ionic composite scrolls and egg-and-dart ornament, supporting the unique carved Scottish Double Rose with the nicks and bruises of time that attest to well over two centuries of sun and storm.

Details repeated in the window ornaments, here stripped of paint, are the Grecian chain or guilloche border, flanked on the east and west by fish-scaling in the small console supports, varied on the north with supports of carved acanthus leaves. Rich moldings frame the window surrounds and form water tables, crowning the thicker and thus protruding heads of the basement masonry. The carving is subtle in scale and motif, almost repressed in the overall bulk of the house.

CHAPTER V

Stone for the Porticoes

BENRY LATROBE PRODUCED this fine watercolor, ink, and wash project for President Jefferson, it is believed, about 1807. It appears to have been part of a set of five, which may in fact have been done as late as 1816, in an unsuccessful attempt to woo the thrifty President Madison into a more ambitious project. Latrobe adapted the designs of the pilasters already present, borrowing the Ionic composite capitals apparently without the Scots' roses. While his series of pictures, intended to show improvements to the house, made the first full reproduction of porticoes, it cannot be used to attribute the porticoes entirely to Latrobe. James Hoban said he had visited the subject with George Washington and later made drawings; Charles Bulfinch adapted Hoban's drawings to finalize the design that was built. All these builders may have had a part. However, it was Hoban who designed and built the South Portico in 1824–26 for President James Monroe, who started the movement for a North Portico, at last realized during the second year of Andrew Jackson's administration.

the East front of the President's House, with the addition of the North and South Porticos

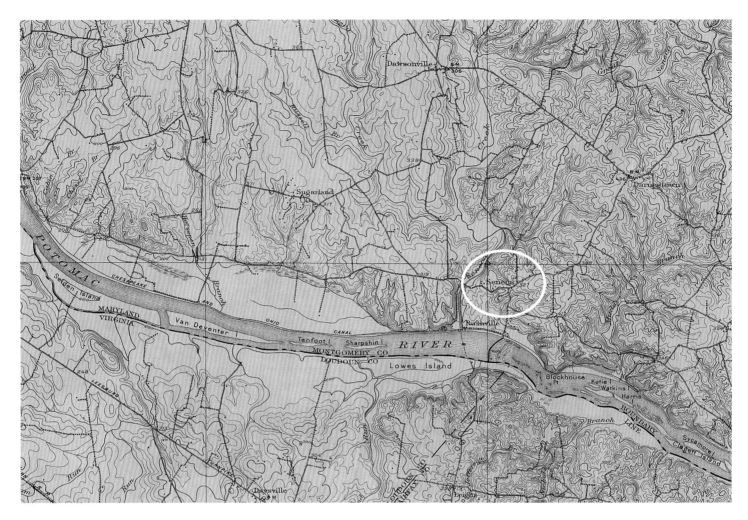

In 1824, when the columned porch was added to the southern bow of the White House, the best of the Aquia stone had been harvested and stone from the quarry upriver at Seneca (encircled on the 1907 topographical map above) was more commonly in use. Also an early quarry, Seneca remained in production long after Aquia, and it produced a much more durable stone. For many years it matched the color of Aquia stone, and it became more convenient to transport, thanks to the C & O Canal, opened in 1830, and the coming of the steam-empowered riverboat, which conquered adverse current. However, in 1824, as the South Portico was being built, the color began to turn a dark, blood-red color. Architect Hoban was unwilling to use the dark Seneca for the columns (even though they were to have been painted), and he rushed back to Aquia to search for a sufficient quantity of lighter colored stone for his portico columns. Today, like Aquia, the Seneca quarry is abandoned. The following recent photographs show what is left, trees and weeds growing into the cracks of remaining outcroppings of rock of various colors. Seneca's remaining outcroppings gained their scars years after the building of the White House porticoes. It was popular in Washington and Baltimore, readily adorning many houses and commercial buildings as trim or full walls.

Although Hoban insisted on Aquia sandstone for the columns of the South Portico, he was willing to use the red Seneca stone on the base. Pictures taken during the conservation of the White House stone, while the paint was stripped away, confirm that Seneca stone was indeed a major part of building the porticoes. A Seneca stone (opposite) in the quarry retains half-impressions of the iron rod used to split the stone, but it could as easily be a split from the 1890s as the 1820s.

Stafford July 1st 1824

Sir

Your letter of the 26th Ulto. was not received untill this evening, With respect to the fissure or vein reported to you, as having appeared crossing the Quarry, It is no recent discovery having been visible ever since I ran my groov at each end of the Column rock, nor is it any thing unusuale in free Stone quarrys, we had come to one before You and Mr. Blagden visited the Quarries last fall, since which we came to a second and third, the two last have been somewhat more oblique than the first, but having a front of forty five feet I am enabled to shaft the column shafts so as to occasion verry little delay or impediment, To give You a more correct view of the situation annexed is a rough sketch of the Quarry showing all the fissures (or Backs as we term them) that have presented themselves. Tho I have run my groves quite as far back as I shall have occasion in pro=cureing all the Column shafts wanting. You may rely on my promptly and particularly reporting to You every materiale change or occurance relative to the Quarry or Column Shafts, I regret that I did not receve Your letter sooner, S. this Morning at 9 o=
=Clock

finished puting two Column shafts on board the Vessel which the Captain told me was all he was to take this load, the balance he said, was to be taken from I. H. Little &c. and moved off amediately. I had another excellent Column quarried and on the Main along side where the vessel lay, but he came prepared to take only two. I have got nearly thro' all my former bills which I hope you will take into consideration as I have a great deal of surplush length of rock, which has to be removed in quarrying the Column Shafts.

I am, Sir, Most respectfully
Your Ob'd Serv't
Tho. Towson

J. Elgar Esq.

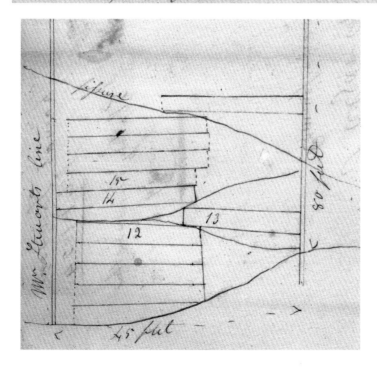

In his letter of July 1, 1824, to Public Buildings Commissioner, Joseph Elgar, Thomas Towson, quarry manager at Wiggington's Island, verified that the quarry held sufficient stone to create the columns of the South Portico. In the accompanying drawing, which shows the outcropping from above (as well as Mr. Steuart's boundary line), Towson shows how the stone will be harvested for the columns and even suggests that more than enough for the necessary six columns shafts was available.

The South Portico as it appears today, from across the South Lawn. The second story balcony was added by President Harry S. Truman in 1948.

Stafford Octr. 19th 1824

Sir

On My return to the Quarry I find sufficient Rock cleared to Make Six Collumn Shafts, Two of which will be ready to Ship by the last of Next week, and two for sortued thereafter, and I am happy to inform You that the quality of the Stone is improving

I am Sir Most Respectfully
Your Obt Servt.
Tho. Towson

J. Elgar Esqr.

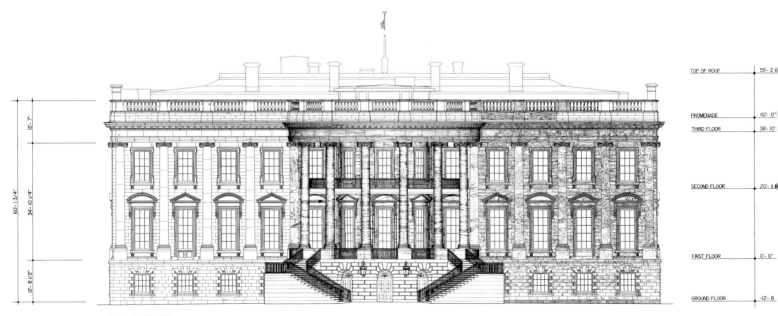

FULL ELEVATION ELEVATION SHOWING GRAIN PATTERN OF AQUIA SANDSTONE
NOTE: STONES WITHOUT GRAIN PATTERN WERE PAINTED OR INACCESSIBLE AT TIME OF RECORDING

SOUTH ELEVATION

In October 1824, Thomas Towson returned to Aquia and reported to Commissioner Elgar (opposite, top) that sufficient rock had been cleared to make the six columns of the South Portico and that the quality of the stone was improving. In 1988–92 the Historic American Building Survey documented the grain pattern of the Aquia sandstone blocks on the south front and portico columns (opposite, bottom). A view from the South Portico (above), looking towards the Washington Monument, captures the monumental scale of the columns, now painted white.

Details on the South Portico reveal the full extent of the stone carving that crowns the White House. Double Scottish Roses and acanthus leaves form column caps; modillions, dentils, and moldings cross the great curve and rejoin the same pattern on the running entablature of the house. A massive stone balustrade bands it all, partially concealing the roof. The balcony was built by President Truman before he began his renovation of the building in 1948. The architect of this innovation, William Adams Delano, made the balcony as "ribbon thin" as modern steel and concrete construction would allow.

A massive three-sided colonnade of stone, the North Portico was added in 1829–30, thirty years after the White House was completed. As on the South Portico, the Double Scottish Rose is a prominent feature of the carving. Planned as an addition even before the fire of 1814, this portico closely resembles that of the Virginia State Capitol, completed 1798 and designed by Thomas Jefferson, making one wonder to what extent Jefferson himself influenced it.

Seen here about 1985, stripped of its white paint during preservation of the stone, the columns of the North Portico (above) show a blush of color suggesting that its construction may combine both Seneca and Aquia stone. The columns are stacked in sections. Once the paint was removed the fine detail of the stone carving was revealed (opposite).

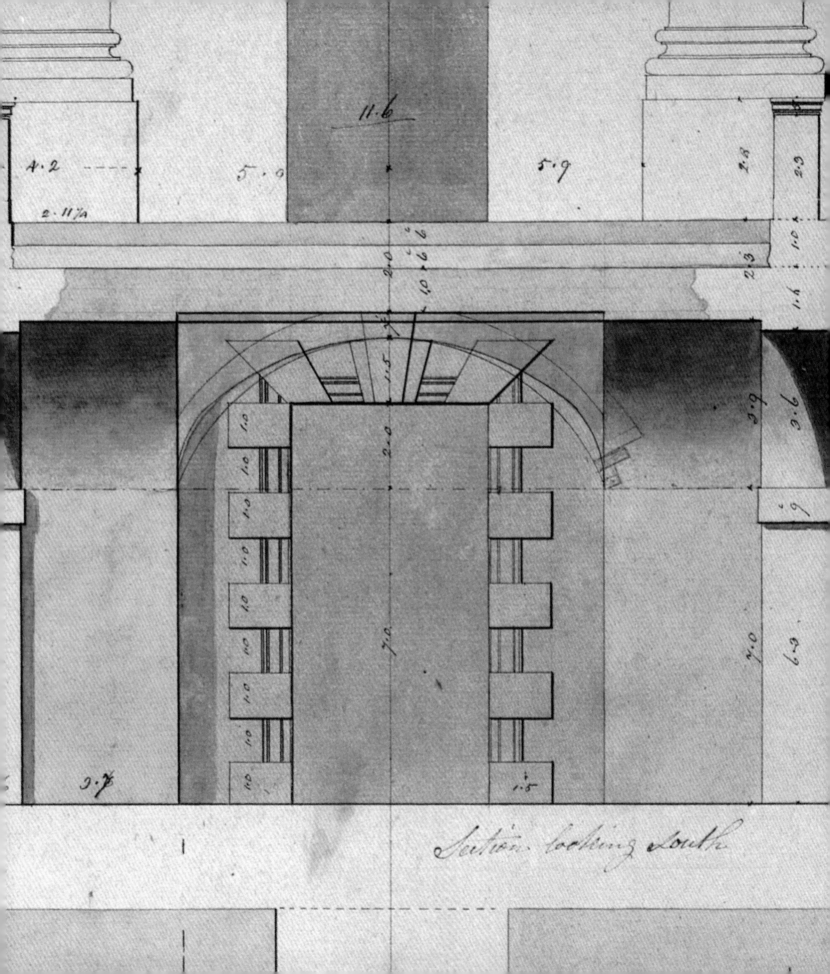

Section looking South

In 1829 when James Hoban commenced the North Portico, he retained Latrobe's north platform and steps. In modifying the previous landing, Latrobe had extended an episode of arching in the interior of the basement to provide a supporting vault beneath the new platform. Latrobe's drawing of the north elevation looking south through the new platform and steps includes Collen Williamson's ornamental basement entrance, with its flanking rustication around the door and keystone. His new vault violates the entirety of the door surround. The door remains in use today (above left). Williamson's design for this north basement door surround very likely came from James Gibbs's Book of Architecture *(1728), probably that century's most referenced guide to classical building. However, in the distance of years between 1728 and the mid-1770s the same elements in Gibbs are also found at Christ Church (above right).*

CHAPTER VI

Survival and Preservation

THE WHITE PAINT that gives the President's House its name has protected the Aquia sandstone for more than two centuries. Presidents had many approaches to keeping the house white. Lincoln painted only the south side, finding the rest white enough. President Grant called in the firemen to hose down the entire outside. In 1798 the first painters used whitewash applied with straw brooms. Actual paint came in 1817 with the purpose of covering dark spots of soot stain left by the fire. Layers of paint built up about every twenty years. During the Carter administration, the National Park Service began a more than twenty-year restoration of the stone, removing the paint entirely and revealing the stone smoothed and carved by stone masons in the 1790s and scarred by fire in 1814. Following twenty years of labor to remove the original paint and repair the stone, the painters returned. Applied in a thin coat, with the use of a spray gun, the new coat of paint resulted in the White House looking the same, but it was protected and conserved for times to come.

123

The White House, burned by British torches in the summer of 1814 during the War of 1812, stood in limbo for only a short time until President James Madison ordered it "repaired." He was very serious. The Scots' walls were laid open. The two ends of the house were pulled down to the level of Collen Williamson's basement, which survived intact with its arching and vaults. The north front was pulled down to some extent, except for the frontispiece and pediment. The entire south front survived. James Hoban, called back to the rebuilding, made use of every stone he could to honor and please President Madison. The window pediment above, when stripped of paint in the late twentieth century, revealed severe scars from the fire and Hoban's repairs for Madison, in which he reused even the smallest stones.

From 1948 to 1952, The White House interior was gutted and rebuilt a second time, and again the exterior stone walls survived (above). This photograph (opposite) captures the exposed old walls with their rubble of stone and brick, newly underpinned with reinforced concrete. (Note the block wall with rectangular openings.) The old basement level is visible above the underpinning. Great heaves of earth are part of the digging for additional cellars beneath the "Ground Floor," as the basement had long been renamed. The steel structure, part of the engineers' plan to hold the house together while it was dismantled, remains intact within the walls of the White House today.

The White House water table, laid originally under the supervision of stonemason Collen Williamson, crowns the basement construction. Above it the thinner walls of the house rise two full stories to a "third floor" attic. The differentiation in thickness of the walls is accompanied by the heavy molding or "water table." Preservation is in process at a basement window (above), showing brick backing to the stone.

During the Truman renovations of 1948–52, the old basement support system of arches and groin vaulting was opened up revealing original groin vault construction in brick and stone made under the supervision of Collen Williamson and Jeremiah Kale. Groin vaulting is here shown cut off from the point of joining the general arching through the basement. This point of joining at a groin gave the heavy house above, by distributing weight, a powerful spine of support. The original construction survived the 1814 burning of the house, but was entirely destroyed in the 1948–52 renovation.

In a project that began with a study in 1976, the National Park Service stripped more than thirty layers of paint from the White House walls. The paint had become so thick that it failed to adhere and had begun to hide the original stone carvers' work. In 1988, James I. McDaniel, then National Park Service associate regional director, reported, "We've uncovered wonderful ornamentation, rosettes, dentil work—stone carving that had been completely obscured by the paint layers. Now they have lights and shadows." After stripping, the porous sandstone was allowed to dry for several months before a fresh thin coat of paint was reapplied.

The removal of some forty layers of paint down to the base whitewash required at junctures ghostly wrapping of the house to keep the misting acids of paint removal from drifting too far away. Above, the South Portico wears its obligatory toga, while the entrance to the house on the south remains open for continued use. Master mason Patrick Plunkett (top and opposite) is seen at work on the project.

CHAPTER VII

Stone in the Monumental City

THE WHITE HOUSE SHARES a part of its architectural history with much of the Washington, D.C., skyline, beginning with the oldest parts of U.S. Capitol, for which the Aquia stone of Government Island was also quarried. The stone of Government Island can still be seen in the city's boundary markers, the U.S. Capitol gatehouses and gateposts, the Old Patent Office Building, and the earliest portion of the U.S. Treasury Building. The blood-red Seneca stone quarried upriver was painted white once set in the base of the South Portico of the White House, but it's familiar color distinguishes James Renwick's romantic Smithsonian Castle. The more durable Vermont and Georgia marbles and granite would eventually replace sandstone in the construction of the city's landmarks. Three different kinds of marble, in fact, were used in the construction of the Washington Monument alone.

The oldest external walls of the United States Capitol were constructed of Aquia sandstone. Completed about 1821, the old Capitol is depicted above as finished—its Aquia stone walls and Corinthian columns painted white and its tar-painted canvas dome cover an endless maintenance challenge. In the mid-nineteenth century, during a massive enlargement of the existing thirty-year old structure ordered by President Millard Fillmore, this dome was replaced with the one we know today. At left is one of B. Henry Latrobe's "corn cob columns" (so-called to the architect's dismay). Carved from Aquia sandstone they survived the burning of the Capitol during the War of 1812. When the deteriorating Aquia stone columns on the east facade of the Capitol no longer could remain, they were taken to the National Arboretum and set up as garden sculpture (opposite).

One of the last monumental works in Aquia sandstone, the U.S. Patent Office was begun during the administration of Andrew Jackson in 1836. The daguerreotype (above), made by John Plumbe in c. 1846, shows the building as it was originally completed of dressed Aquia stone. The upper floors succumbed to a fire in 1877, but the surviving outer walls of the first floor can still be seen today in the covered courtyard shared by the Smithsonian Institution's American Art Museum and National Portrait Gallery. Designated a historic landmark in 1965, the building has housed the two museums since 1968.

Aquia sandstone was used to build these cenotaphs ordered by Congress for the Congressional Cemetery near the Capitol in Washington. Designed in 1807 by B. Henry Latrobe, the vaults were intended for the remains of statesmen who died while on duty in Washington. Some bodies stayed while others were eventually moved to their home states, leaving their cenotaphs empty. The cenotaph holding the remains of Dr. William Thornton, a member of the commission that built the White House and a great friend to the building men, is pictured above.

The surviving Aquia and Seneca stone landmarks are so much a part of the historic fabric of the city that one might not notice them. Thousands of commuters and tourists making their way through the busy intersection at Constitution and Fifteenth Street (opposite) each day pass a weathered gatepost and guard house that stood on the grounds of the U.S. Capitol from c. 1827 to 1874. Two miles away, across the river, thousands of visitors to Arlington National Cemetery each year pass beneath the thirty-foot high McClellan Gate (above), constructed of Seneca sandstone in the 1870s.

Seneca stone was selected over many other types of sandstone and marble by the Smithsonian Institution's building committee for the distinctive exterior walls of the original "Gothick" museum. The committee determined that the red Seneca stone would be durable and more appropriate than marble for the architectural style adapted by James Renwick Jr., the architect. Renwick himself visited the quarry to select the stone. Now known as the "The Castle," the building was completed in 1855.

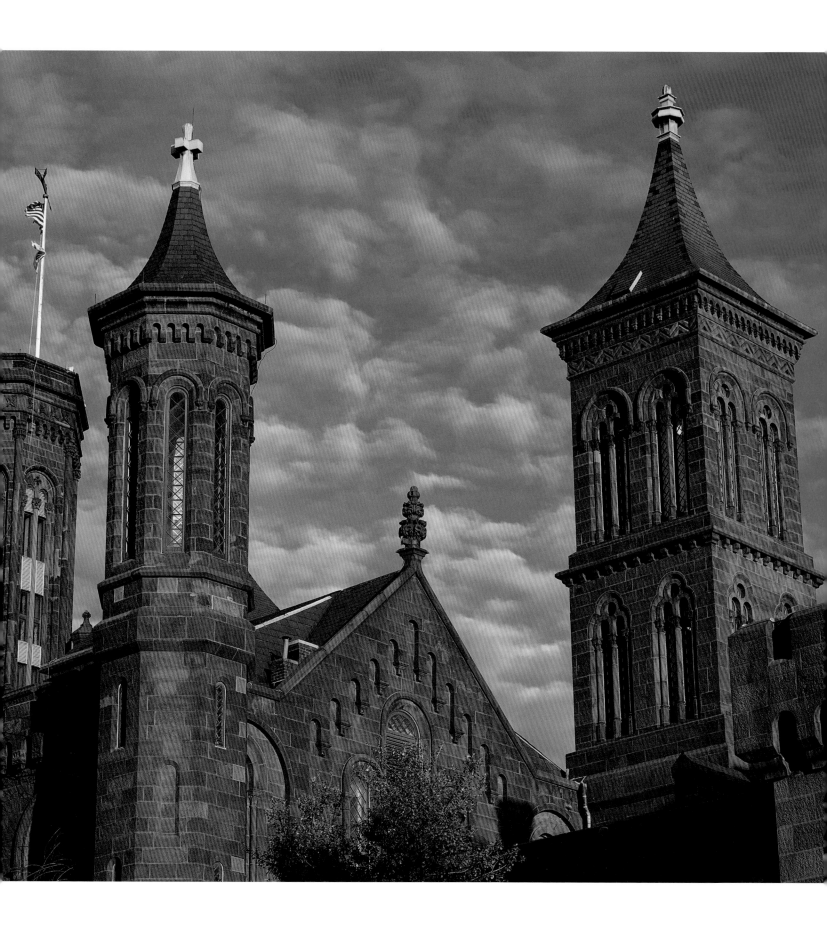

The grave of John (known as Jack) Clipper, a freeman and stonecutter, is marked with Seneca sandstone (opposite). It is one of five hand-carved headstones remaining in the quarry cemetery. Born into slavery in 1840 in Hanover, Virginia, Clipper was freed by the Union Army during the Civil War and made his way to Maryland where he worked at Seneca quarry mill and lived in a house close to the C&O Canal with his wife and ten children. The lockkeeper's house (above) was built of Seneca sandstone about 1829, just as the North Portico was being built.

Reflections

WHY A BOOK about the Scots stonemasons? Principally because their stonework is all that is left of the original White House conceived by Congress and ordered built by George Washington. Further, it is interesting to get to know people largely forgotten, whose hands created in the dawning years of the nation an enduring monument that, for all its vicissitudes, remains strong and stable, and is central to American life today.

History, first the story of human feelings and activities, requires a context. History is shaped within the context because the context molds and influences the players. The history of the White House begins with that one 1790 act of Congress, simple enough in print. The rest of the story expands, including George Washington's efforts and urgencies, as the act is carried out. The power sparked by the Residence Act drew hundreds of people to effect its realization, thus hundreds of stories, hundreds of circumstances. Records of most of them vanished.

The Scots were swept into that flow, carried first from the Highlands, town of Dyke, the country house of Moy, then from the sophisticated building project creating New Town, Edinburgh. They came to America to build a big house of stone. No record gives us their opinion or describes the pride that must have come with building a house for George Washington, the unblemished world hero of their time. We know little of what they thought except what we can surmise by placing them in the details of the life that surrounded them. We see what they built. They performed their work and, when they departed for home, disappeared from the context of the White House forever, presumably keeping their American experiences alive for awhile in tales they told fireside on Scottish nights.

Illustration Credits

All illustrations in this book are copyrighted as listed below and may not be reproduced without permission of the copyright owner.

AP–Associated Press Photo
HABS–Historic American Building Survey
LOC–Library of Congress
MDHS–Maryland Historical Society
NARA–National Archives and Records Administration Collection
NPS–National Park Service
OCWH–Office of the Curator, The White House
WH–Official White House Photo
WHHA–White House Historical Association
WHHA / WH Collection– Copyright White House Historical Association / White House Collection

II–XIV	Bruce M. White for WHHA
20–25	Dahl Taylor for WHHA
28–31	Dahl Taylor for WHHA
47	Richard Cheek for OCWH
51	Bruce M. White for WHHA
53	LOC
54–55	MDHS
55	*Top:* WHHA / WH Collection
55	*Bottom:* Martin Radigan for WHHA
56–69	Martin Radigan for WHHA
70–73	Bruce M. White for WHHA
74–75	WHHA
77	*Top:* MDHS
77	*Bottom:* Bruce M. White for WHHA / NARA
78–81	Bruce M. White for WHHA / NARA
83	WHHA, NARA
84	Jack E. Boucher, HABS, LOC
85	OCWH
87	HABS, LOC
88–91	Bruce M. White for WHHA
92	Oak Spring Garden Library
93–96	Bruce M. White for WHHA
97	Martin Radigan for WHHA
98–99	Bruce M. White for WHHA

101	LOC	132–3	Denis Paquin, AP
102	U.S. Geological Survey	133	Jack L. Boucher, HABS, LOC
103–6	Martin Radigan for WHHA	135	Martin Radigan for WHHA
107	Erik Kvalsvik for OCWH	136	*Top:* LOC *Bottom:* Architect of the Capitol
108–9	Bruce M. White for WHHA / NARA	137	Martin Radigan for WHHA
110–11	Bruce M. White for WHHA	138	LOC
112	*Top:* Bruce M. White for WHHA / NARA	139	WHHA
112	*Bottom:* HABS, LOC	140–1	Tim Evanson
113–7	Bruce M. White for WHHA	141	Ted Nigrelli
118	Jack E. Boucher, HABS, LOC	142	WHHA
119	Richard Cheek for OCWH	143–7	Martin Radigan for WHHA
120	LOC	149	WHHA
121	Bruce M. White for WHHA	154	Bruce M. White for WHHA
123	OCWH		
124	WHHA / WH Collection		
125	Erik Kvalsvik, WH		
126–7	Abbie Rowe, NPS, Truman Library		
128	OCWH		
129	*Top:* Abbie Rowe, NPS, WH		
129	*Bottom:* Dahl Taylor for WHHA		
130–1	Jack E. Boucher, HABS, LOC		

Author

WILLIAM SEALE is an American historian and author. He attended Southwestern University in Texas and completed his Ph.D. at Duke University in North Carolina. An independent scholar since 1965, he has written extensively on the White House and has participated in the restoration of many state capitols. His many books include the *Imperial Season* (2013); *Blair House: The President's Guest House* (2016); and the two-volume *The President's House* (1986 and 2008). He is editor of the journal *White House History*, the award winning quarterly of the White House Historical Association.

Principal Photographers

MARTIN RADIGAN has spent more than a decade seeking out and photographing beautiful places—some far away, some not. Landscape has always been the favorite subject and primary focus of his work. Martin also enjoys photographing people and wildlife and once, while lost in the moment of shooting, he found himself on opposite sides of a car from a Grizzly Bear (of course he lived to tell about it and the images were pretty good). Radigan's work has appeared in a wide range of publications, including those of the White House Historical Association.

BRUCE M. WHITE is a fine art and architectural photographer. Formerly a staff photographer at the Metropolitan Museum of Art and Sotheby's, White now works on commission for museums, artists, art collectors, universities, and publishers in the United States and abroad. For the White House Historical Association, White authored *At Home in the President's Neighborhood: A Photographic Tour* and served as principal photographer for *The White House: Its Historic Furnishings and First Families*, *The White House: An Historic Guide*, and many other titles.